T0195312

An Equation for Every Occasion

An Equation for Every Occasion

Fifty-Two Formulas and Why They Matter

John M. Henshaw

With contributions from Steven Lewis

Johns Hopkins University Press

Baltimore

Printed in the United States of America on acid-free paper
2 4 6 8 9 7 5 3

Johns Hopkins University Press
2715 North Charles Street
Baltimore, Maryland 21218-4363
www.press.jhu.edu

Library of Congress Cataloging-in-Publication Data

Henshaw, John M., author.
An equation for every occasion : fifty-two formulas and why they matter / by John M. Henshaw ; with contributions from Steven Lewis.
pages cm
Includes bibliographical references and index.
ISBN-13: 978-1-4214-1491-1 (hardcover : alk. paper)
ISBN-10: 1-4214-1491-0 (hardcover : alk. paper)
ISBN-13: 978-1-4214-1492-8 (electronic)
ISBN-10: 1-4214-1492-9 (electronic)
1. Mathematical analysis. 2. Mathematical models–Miscellanea. I. Title.
QA401.H457 2014
510—dc23 2013050140

A catalog record for this book is available from the British Library.

Special discounts are available for bulk purchases of this book. For more information, please contact Special Sales at 410-516-6936 or specialsales@press.jhu.edu.

Johns Hopkins University Press uses environmentally friendly book materials, including recycled text paper that is composed of at least 30 percent post-consumer waste, whenever possible.

To my students at the University of Tulsa

Contents

Preface ix

1. As the Earth Draws the Apple 1
2. And All the Children Are Above Average 4
3. The Lady with the Mystic Smile 8
4. The Heart Has Its Reasons 11
5. AC/DC 14
6. The Doppler Effect 18
7. Do I Look Fat in These Jeans? 22
8. Zeros and Ones 25
9. Tsunami 28
10. When the Chips Are Down 31
11. A Stretch of the Imagination 34
12. Woodstock Nation 37
13. What Is π? 40
14. No Sweat 43
15. Road Range 46
16. The Bends 50
17. It's Not the Heat, It's the Humidity 54
18. The World's Most Beautiful Equation 58
19. Breaking the Law 61
20. The Mars Curse 64
21. Eureka! 67
22. A Penny Saved . . . 70
23. If I Only Had a Brain 73
24. Because It Was There 77

25. Four Eyes 80
26. Bee Sting 83
27. Here Comes the Sun 87
28. A Leg to Stand On 90
29. Love Is a Roller Coaster 93
30. Loss Factor 96
31. A Slippery Slope 100
32. Transformers 103
33. A House of Cards 107
34. Let There Be Light 110
35. Smarty Pants 113
36. As Old as the Hills 116
37. Can You Hear Me Now? 119
38. Decay Heat 123
39. Zero, One, Infinity 127
40. Terminal Velocity 130
41. Water, Water, Everywhere 133
42. Dog Days 136
43. Body Heat 139
44. Red Hot 143
45. A Bolt from the Blue 146
46. Like Oil and Water 149
47. Fish Story 153
48. Making Waves 156
49. A Drop in the Bucket 160
50. Fracking Unbelievable 164
51. Take Two Aspirins and Call Me in the Morning 167
52. The World's Most Famous Equation 171

Bibliography 175
Index 185

Preface

This is not a math book. It's a storybook. It's a book of fifty-two true stories, and as it happens, each one of those stories was inspired by a mathematical equation. Behind every equation lies a story—sometimes many stories. And behind many stories, there lies an equation—sometimes more than one. Our goal was to find those stories, and those equations, and to tell them as well as we could.

As a result, we are unable to avail ourselves of the sage advice given to the incomparable Stephen Hawking when he wrote the classic *A Brief History of Time*. In the acknowledgments to that book, Hawking notes that he had been warned that each equation he included in his book would halve the book's sales. He thus decided to include only one such potential revenue reducer, Einstein's famous equation,

$$E = mc^2.$$

While our book is also written for a general audience, its purpose is to present a new equation at the beginning of every chapter. It is thus our fervent hope that sales of this book do not adhere to the relationship suggested by Professor Hawking's advisor, presented here in the form of an equation:

$$\text{Actual sales} = \frac{\text{Potential sales}}{2^n},$$

where n is the number of equations in the book.

There comes a point in the education of many children, particularly if they like math, when they realize that math is not just another subject to be studied and compartmentalized along with all the others. Often, it is in the early teenage years that many kids begin to realize that math is an indispensable tool for explaining the how and the why of so many things. A little bit later, those same teens begin to discover that math is good for something else, and that is for making the world a better place. Somewhere along the way it occurred to us that math has yet another purpose, and that is in storytelling. The stories in this book come from the worlds of science and engineering, but also from the worlds of business, the arts, and recreation. A good equation, like a good story, is where you find it.

In *How to Read a Book*, Mortimer J. Adler and Charles Van Doren offer sound advice on many things, including reading books that are liberally sprinkled with equations. They suggest, in part, that "skipping is often the better part of valor." We agree. We hope that you will skip around lots as you read this book. The stories are short—one thousand words each, give or take—and they can be read in any order that you desire. Many, if not most, of the equations we write about merit longer treatment. Entire books—in some cases, shelves full of books—have been written about some of them. That is not our purpose here. If, having read our book, you find yourself wanting more, we hope you will seek out some of the references we have provided for each story.

A mathematician, according to Lord Kelvin, is someone to whom the fact that the integral, from minus infinity to infinity, of $e^{-x^2}dx$ is equal to the square root of pi is as obvious as it is to you and to us that two plus two equals four. In equation form:

$$\int_{-\infty}^{+\infty} e^{-x^2}\,dx = \sqrt{\pi} \qquad\qquad 2+2=4.$$

Lord Kelvin was right about most things (with a few famous exceptions), and we suspect his definition of a mathematician is not far from the mark. We hasten to add that we wrote this book for those of us in the latter (two plus two equals four) crowd. To the extent that those few members of the former (square root of pi) group find it an interesting read as well, we shall be grateful.

This book had its origins in an innocent little conversation I had with one of my students a few years ago. That was the day that Steven Lewis stopped by my office to see if I might be interested in pursuing an independent study project with him. "Aren't you the guy who is double majoring in mechanical engineering and English literature?" I asked. He replied that this was indeed the case. "Well, then, maybe we should write a book together!" Although my offer was made somewhat in jest, it was eagerly accepted. And so, each week for the rest of the semester, we wrote one "equation story" apiece. During our weekly meetings we read our stories to each other and critiqued and discussed each other's work. It was certainly the most unusual, and one of the most enjoyable, independent study projects I've ever pursued. Steven eventually graduated and began his career as an engineer. And so it fell to me to round out the number of stories to fifty-two, do some editing, and wrap things up. But I shall always be grateful that Steven, with an undergraduate's optimism, was so willing and able to help me get this project off the ground.

An Equation for Every Occasion

1. As the Earth Draws the Apple

$$F = G\frac{m_1 m_2}{r^2}$$

Newton's Law of Universal Gravitation

Newton's law of universal gravitation, which appears in Isaac Newton's *Principia* and was first published in 1687, states that the force, F, acting between two bodies of masses m_1 and m_2, is directly proportional to the product of the masses ($m_1 \times m_2$) and inversely proportional to the square of the distance, r, between the bodies. In a consideration of the gravitational force of a large body such as a planet, the mass of the planet is taken to be concentrated at a single point at the very center of the planet. The gravitational effects of a spherically symmetric mass distribution, in regions exterior to it, is indistinguishable from that of a point mass at the center of the sphere. G is the universal gravitational constant, which is approximately 6.67×10^{-11} m^3 / (kg · s^2).

As the famous trees of history go, Isaac Newton's apple tree ranks right up there with the one a young George Washington supposedly copped to having chopped down. But Newton didn't go after his apple tree with an ax; he merely observed it releasing an apple, which descended straight down to the ground below it. While there is no evidence that the most famous piece of fruit in the history of science actually struck Newton on the head, the rest of the story is probably true. In his ruminations on the nature of what we now call gravity, Newton was indeed inspired by a falling apple. The eventual result was our equation here, called Newton's law of universal gravitation.

Isaac Newton (1642–1727) was perhaps the greatest and most important scientist who ever lived. Just reading an abridged list of his monumental accomplishments can wear you out. These include, in addition to the subject of this story, the basic laws of optics, the three great laws of motion (Newton's first, second, and third laws), important advances in thermodynamics (including Newton's law of cooling), and contributions to pure mathematics (such as the development of calculus) so valuable that Newton is almost universally acclaimed as one of the three or four greatest mathematicians who ever lived. Newton's magnum opus, *Philosophiae Naturalis Principia Mathematica* (*Mathematical Principles of Natural Philosophy*), was published in 1687 and is still hailed as one of the most important scientific works ever published.

Newton himself repeated the falling apple story on numerous occasions, citing it as being among his inspirations for understanding what was really on his mind, which was a better understanding of the motions of the Earth, the moon, and the other planets in our solar system. Newton realized that just as the falling apple was drawn by the Earth, the Earth was drawn by the apple. But the mass of the apple is negligible compared to that of the Earth, and the effect of the apple on the Earth is likewise negligible. Not so with the Earth and the moon, however. The Earth has about 80 times the mass of the moon, but the moon is no apple. Its mass is plenty large enough to measurably influence the motion of the Earth.

Gravity is a mysterious force, and Newton's law of universal gravitation was by no means the last word on this subject (as we shall see). Gravity on the planet Earth was empirically well understood in Newton's day. Apples did fall straight down, after all. But folks had a hard time believing that that the same force that caused apples to fall from trees could explain the orbits of planets.

But that's just what the law of universal gravitation does. Using his law, Newton was able to accurately predict the motion of the moon around the Earth, among other things. But it wasn't until well after Newton's death in 1727 that perhaps the most important experimental confirmation of the law was made. The planet Uranus was discovered by the astronomer William Herschel in 1781. By 1846, Uranus had nearly completed one single orbit around the sun since its discovery. But this had given astronomers enough time to discover several anomalies in the orbit of Uranus that could not be explained by the law of universal gravitation. Unless, that is, there was another planet hiding somewhere out there in the solar system, another planet whose mass was causing the unexplained characteristics of the orbit of Uranus.

It fell to a mathematician, not an astronomer, to discover that planet: Neptune. In 1846, the French mathematician Urbain Le Verrier used the law of universal gravitation to correctly predict the location of Neptune, which was soon thereafter verified by astronomers in Germany. Le Verrier thus discovered a planet not with a telescope, but with "the point of his pen," as was famously remarked at the time by the physicist François Arago. Although Newton had been dead for well over a hundred years, his law of universal gravitation was finally confirmed.

Or was it? While he was analyzing the orbit of Uranus, Le Verrier was also trying to explain mathematically some tiny, nearly imperceptible perturbations in the orbit of the planet Mercury that had been discovered by astronomers. Le Verrier theorized that these too might be accounted for by an as yet undiscovered planet. The undiscovered planet was even tentatively named Vulcan. But Vulcan was never found, and in fact it does not exist. Other explanations, in accordance with Newton's law, were proposed, but all were found wanting.

In 1916, Albert Einstein proposed three tests for his general theory of relativity. One of these was the very problem Le Verrier had struggled with seventy years before: the unexplained perturbations in Mercury's orbit. Einstein's theory of general relatively describes gravitation in terms of the curvature of space-time. The theory is indeed general, applying to everything from black holes to planets to Newton's apple. The departures (the difference) between Newtonian gravitation theory and general relativity are usually negligible for planets, but for black holes and neutron stars those departures are dominant. And thus the "anomalous precession" of the orbit of Mercury, the closest planet to our massive sun, cannot be calculated from Newton's law. Mercury's deviations from Newtonian behavior are tiny, but they are there. Einstein's theory of general relativity does explain the anomalous motion of Mercury, however, and when Einstein demonstrated this in 1916, it was powerful evidence that his theory was correct.

All of this does nothing to diminish the contributions of Newton, in this or any other area. In a letter to his rival Robert Hooke, Newton famously remarked, "If I have seen further, it is by standing on the shoulders of giants." Even today, Newton remains a giant whose shoulders continue to support the scientific work of legions.

2. And All the Children Are Above Average

$$\phi(x) = \frac{1}{\sqrt{2\pi}} e^{\frac{-x^2}{2}}$$

The Standard Normal Distribution, or "Bell Curve"

The probability distribution function, $\phi(x)$, is shown for a distribution with a mean value of zero and a standard deviation of one. When $\phi(x)$ is plotted on the vertical axis of a graph versus x on the horizontal axis, the familiar bell curve results. The peak of the curve, or maximum value for $\phi(x)$, occurs, in this case, at $x = 0$. The curve describes the statistical distribution of data that are said to be "normally distributed." The form of the equation is a little more complicated if the mean value is something other than zero and/or the standard deviation is not 1. The size of the standard deviation determines how narrow or broad the bell curve is. The larger the standard deviation, the broader the curve becomes.

What are the odds that a basketball player will score twice as many points as his average in a single game? How likely is it that a baby born today will live to be 100 years old? If certain markers in a suspect's DNA match those found at the crime scene, what are the chances that the DNA was left there by someone else?

Questions like these lie in the realm of probability theory. The probabilities of all sorts of things are described by our subject here, the bell curve. The bell curve is variously known as the normal distribution, the standard distribution, and the Gauss curve, among other things. What does it mean when we say that a certain set of data or information follows a bell curve? Well, consider the average American adult male, who is about five feet ten inches tall. Not all American men are five foot ten: some are much shorter or much taller. But if you make a graph with height on the horizontal axis and the

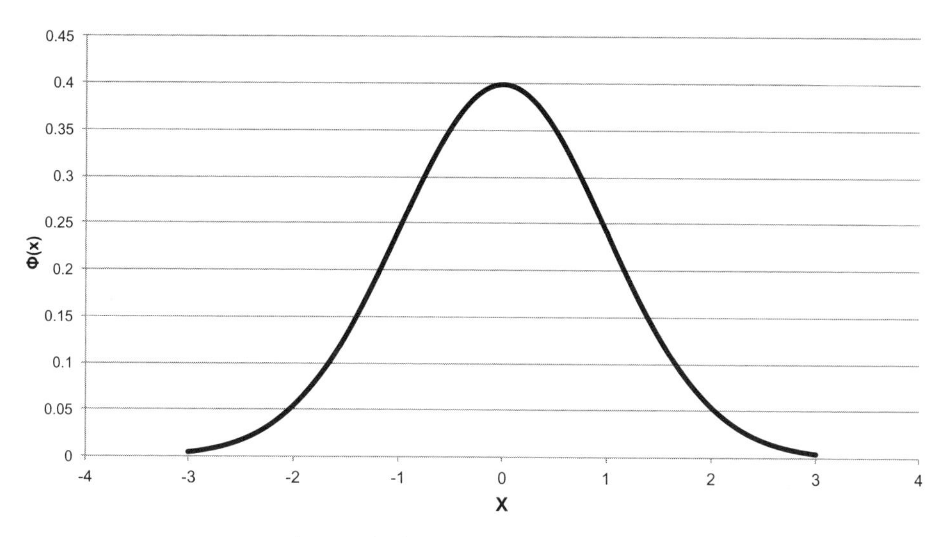

Figure 1. The standard normal distribution, or "bell curve"

number of American men that have any given height on the vertical axis, the graph tends to follow a bell-shaped curve.

To draw a bell curve such as the one shown in figure 1, for height or anything else, you need to know only two things: the average (or mean) value of what you are graphing and the standard deviation. The average value tells you where the peak of the bell curve is. The standard deviation tells you how gently or sharply the curve slopes away from that peak on either side. Our equation above describes a bell curve for the "standard normal distribution," where the average value is zero and the standard deviation is 1. Returning to our height example, the average American man is about five foot ten, and the standard deviation is about three inches. With just those two numbers, we know a great deal about the heights of American men. The bell curve tells us, for example, that just over two-thirds (68.2%) of all American men are between five foot seven and six foot one. In statistical terms, this is just the average ±1 standard deviation, which *always* contains 68.2% of the area under the bell curve.

If about two-thirds of American men are between five foot seven and six foot one, that means that one-third aren't. Half of those men—one-sixth of the total, or about 16 million men—are shorter than five foot seven, and the other one-sixth are taller than six foot one. The odds drop off fast the farther you get from the average value. As any basketball fan can tell you, there are lots of players six foot ten or taller. But six foot ten is 12 inches, or four standard deviations, above the average. In the American population of about 100 million adult men, only about 3,200 are six foot ten or taller. If you selected 31,000 American men at random, you would expect to find only one that was six foot ten or taller.

The bell curve comes up a lot in education. Students, for example, often ask if a test or a course will be graded "on a curve." Grading a test on a curve is widely misunderstood, by students and teachers alike. Grading on a curve is a means of relative grading, as opposed to grading on an absolute scale. Do you remember the written test you took before you got your driver's license? Let's say your state requires a score of 70 to pass that test. This is an absolute standard. If you make a 69, there will be no driver's license for you. But what if the state graded this test on a curve? Instead of an absolute standard of 70 to pass, what if the state requires only that your score be above the average of all those who took the test on the same day? If that average is a 62, and you make a 69, then you pass. But if the average is above 69, then you fail. The standard for passing the test is no longer absolute but is instead dependent on the performance of others.

When a pass/fail test is graded on an absolute scale, it is possible that everyone will pass or, conversely, that everyone will fail. But when a pass/fail test is graded on a curve, there will always be some who pass (everyone above the average) and some who fail (everyone below the average). When a test that is assigned letter grades (A, B, C, D, F) is graded on a curve, the same sort of relative grading applies. Those who score near the average get a C, and the other letter grades are distributed around the average, whatever that might be. Thus, for a test that is strictly graded on a curve, there will always be as many F's as there are A's, a fact of curve-graded tests that many students would rather not hear about. In Garrison Keillor's Lake Wobegon, it may be true that "all of the children are above average," but when a test is graded on a curve, this is most emphatically not the case.

The bell curve is without question the most important way in which data are statistically distributed. So many things tend to naturally follow a bell-shaped distribution curve that it's tempting to think that everything does so. But it isn't true; not all data follow a bell curve. For example, the lifetimes of various types of components—say, a cooling fan in a computer—tend to follow a non-bell-shaped distribution. In the case of the lifetime of the cooling fan, imagine a distorted bell curve where the data are skewed towards shorter lives, with a smaller "tail" of very long-lived components.

What difference does it make whether certain types of data follow a bell curve or not? Well, it can make a huge difference, depending on the question. When researchers evaluate the effectiveness of a new drug for lowering blood pressure, for example, the drug is often tested in a clinical trial where some people get the drug and others get a placebo. When the results are later evaluated, knowledge of the statistical distribution of the results is extremely important. If the data are "normally distributed"—that is, if they follow a bell curve—they can be evaluated using techniques that would not be applicable if the data do not fit a bell curve. The question of whether the drug is judged to be effective is closely tied to how the data are distributed.

The mathematics of the bell curve was first published in 1733 by Abraham de Moivre (1667–1754). Later on, a whole slew of heavyweights like LaPlace, Gauss, Galton, Maxwell, and others were instrumental in the development and application of this vital concept from the world of probability and statistics; and their names, particularly that of Gauss, are better known today in this context. That de Moivre's name has largely been forgotten in all this is thus hardly surprising. A master of games of chance, it's likely he would have even predicted it himself.

3. The Lady with the Mystic Smile

$$\frac{a+b}{a}=\frac{a}{b}=\frac{1+\sqrt{5}}{2}=\phi$$

The Golden Ratio, ϕ

A rectangle whose long and short sides are a and b, respectively, is a so-called "golden rectangle" if the ratio of a to b is as shown above—approximately 1.618.

What do the dimensions of the Parthenon, the distances between successively smaller tree branches, the proportions of the face of Mona Lisa, and the geometry of a DNA molecule have in common? All contain the "golden ratio," a geometric proportion both beautiful and mysterious. Or do they really? Is this a case of the mathematical precision of nature and aesthetics, or simply wishful thinking on the part of mathematicians and others who've discovered a clever formula?

First things first. The golden ratio is often given the symbol ϕ, the Greek letter phi, and is equal to $(1+\sqrt{5})/2$, or approximately 1.618. If you draw a rectangle whose sides are in this proportion—that is, if the longer side is about 1.618 times the shorter one—you have a golden rectangle. A rectangle whose sides are in the ratio of 5 to 8 is very close (within about 1%) to a golden rectangle.

The golden ratio, and thus the golden rectangle, have almost magical mathematical properties. For example, if you subtract a square that is a by a from an a by b golden rectangle, the remaining smaller rectangle (which is $a-b$ by a) is *also* a golden rectangle. If you subtract a square from this new rectangle, the remainder is again golden. And so on and on, forever.

Constructing a golden rectangle is simple. As shown in figure 2, you can do it in three steps with a ruler and a compass. Start with a square (step 1). In step 2, draw a line connecting the midpoint of one side to one of the opposite corners. Let that line be the radius of an arc that you draw (step 3) with the compass. Where that arc intersects a line extended from the side of the original square is the corner of the golden rectangle.

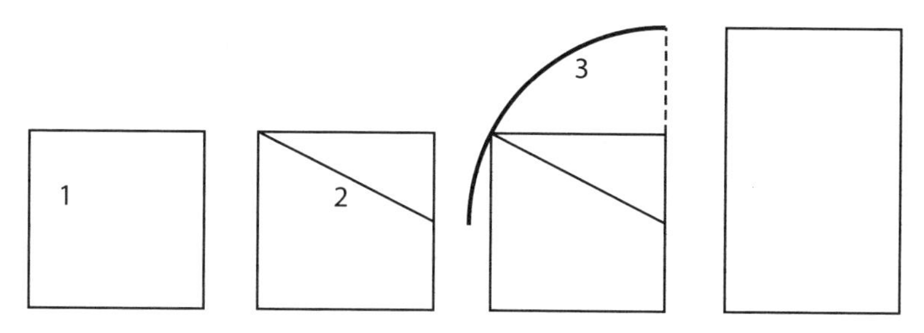

Figure 2. Constructing a golden rectangle from a square

The golden ratio has various other interesting mathematical properties. In fact, entire books have been written on the math involved here. Mathematical interest in what we now call the golden ratio can be traced back at least to the venerable Pythagoras. Euclid later provided the first written definition.

Much later, the golden ratio began showing up in other ways, and it is in these nonmathematical contexts that it is perhaps best known today. Did Leonardo da Vinci, for example, proportion various aspects of the *Mona Lisa* in accordance with the golden ratio?

In the 1860s, the German physicist and psychologist Gustav Fechner, the father of the modern science of psychophysics, conducted perhaps the first attempt to quantify the special aesthetic attraction of the golden rectangle. He asked individuals to select the most beautiful rectangle from among a collection of ten. The proportions of the rectangles varied from square to very tall and narrow (or short and wide, depending on the orientation). Just over three-fourths of the subjects choose either the golden rectangle or one of the two rectangles closest in proportions to the golden rectangle. Combined, the other seven rectangles, which represented the two extremes (square-like or tall and narrow) were selected by fewer than 25% of the subjects of the test.

Fechner's experiment either proves the inherent beauty of the golden rectangle or perhaps simply confirms that people tend to eschew the extremes when given choices that lie along a spectrum. Since Fechner's time, numerous attempts to prove or disprove his conclusions have mostly been negative. Researchers have generally concluded that there is nothing special about the aesthetics of the golden rectangle.

Numerous studies have claimed to find the golden ratio or rectangle hiding out in all kinds of places, such as the drawings and paintings of Leonardo da Vinci or the dimensions of architectural wonders like the Parthenon or the Great Pyramid of Egypt. Subsequent researchers have decided more often than not that there is no compelling reason to conclude that Leonardo or any of these architects either consciously or unconsciously used the golden ratio in their works. It may be that if you search long enough and hard enough for a certain geometric pattern in a painting or a building, you may find *something*—some set of dimensions or ratios—that approximates what you are looking for. Those bent on finding such patterns are aided by the fact no measurements can ever be exact, thus providing a bit of wiggle room when calculating ratios and looking for patterns.

If efforts to find the golden ratio in art and architecture have proven elusive, what about nature? If you divide a person's overall height by the height of his or her navel, the result is frequently not far from the golden ratio. Similar constructs have been alleged for birds, insects, and other creatures. It appears that such studies suffer from the same problem noted above. Measurements are approximate, and the form of a living creature provides an almost inexhaustible supply of dimensions and ratios. Look long and hard, and you will probably find the golden ratio, or whatever else you may be searching for, somewhere.

4. The Heart Has Its Reasons

$$\frac{1}{2}\rho V^2 + \rho g z + P = \text{constant}$$

Bernoulli's Equation for the Flow of an Incompressible Fluid

The velocity of the fluid, at some given point, is V; z is the height of that point; and P is the pressure at that point. The values of V, z, and P vary from point to point within the fluid, whereas the density, ρ, is constant: the equation is limited to incompressible fluids. (The constant g is the acceleration due to gravity.) Bernoulli's equation is thus applicable to liquids, which are essentially incompressible. The equation may be applied to (compressible) gases only in limited cases, such as when the Mach number (the velocity relative to the velocity of sound in the fluid) is less than 0.3.

Engineering professors, when teaching the subject of fluid mechanics, have been known to pose the following question to their students: Which land animal has the largest heart? The most common guess (and the one the professor is hoping for) is the elephant. Having thus trapped his or her students, our erudite friend responds with a flourish that it is not the elephant that possesses the largest heart, but rather the giraffe. It makes for a good story, and there are some sound fluid mechanics reasons why it might be true—but it isn't. The myth of the size of the giraffe's heart is probably based on some old, incomplete data, and it's a story to which we'll return a little later.

But first, let's note that while the giraffe may not have the largest heart, its circulatory system does contain a number of other specialized features, all designed to allow an animal with such unusual dimensions to survive. And so, as a replacement for the "largest heart" question, we offer the following fluid mechanics teaser: What do a giraffe and a fighter pilot have in common? They both wear a g-suit so they won't pass out during maneuvers. The giraffe's g-suit is built in; the pilot's, however, is human-made.

Daniel Bernoulli would not have been surprised by the giraffe's g-suit, or by the other specialized features of the circulatory system in the world's tallest animal. The Swiss scientist and mathematician (1700–1782) did pioneering and extremely important work in several areas, perhaps most notably the field of fluid mechanics. The heart, after all, is just a pump. And the liquid it pumps, blood, obeys the principles Bernoulli elucidated over two hundred years ago.

From a fluid mechanics perspective, all the problems of being a giraffe stem from the great height of this graceful herbivore. Adult giraffes generally range from 14 to 17 feet tall. The average female weighs 1,800 pounds, while the average male tips the scales at 2,600. The myth of the size of the giraffe's heart probably dates back to some sketchy data from the mid-1900s. A 2009 study, based on the analysis of 56 giraffes, found that the giraffe's heart weighs on average about 0.5% of its body weight. This is roughly the same percentage as for most other mammals. However, while the giraffe's heart may not be abnormally large, it is abnormally thick. The wall of the left heart ventricle in an adult giraffe can be over three inches thick! This is much, much thicker than would be expected for an animal of the giraffe's weight. And there's good reason for it.

Perhaps the most important job of the heart (giraffe, human, or other) is to maintain an adequate supply of blood to the brain. If there isn't enough blood carrying oxygen to your brain, you will pass out—very quickly. Just about everyone is familiar with the feeling of light-headedness that can occur if you stand up too fast after being seated for a long time. The rapid upward motion of your head makes it difficult—just for an instant—for the heart to supply enough oxygenated blood to the brain. As a result, your brain is ever so slightly impaired, and you feel lightheaded. A more extreme example is when a fighter pilot executes a severe maneuver in his or her plane, resulting in a longer, more profound lack of oxygen in the brain. This has caused pilots to black out, sometimes with catastrophic consequences.

Hence the development of the jet fighter pilot's g-suit (the *g* stands for gravity). A g-suit is a set of tight-fitting trousers fitted with inflatable bladders. When the plane undergoes a large acceleration, sensors in the suit rapidly inflate the bladders, creating pressure on the pilot's abdomen and legs. This extra pressure reduces the tendency of the blood to pool in the lower body during high g-force maneuvers and is very effective at preventing loss of consciousness. The great distance between the giraffe's head and its lower legs necessitates a similar sort of mechanism. The lower legs of the giraffe are thus covered with a very thick, tight-fitting skin that acts much like a pilot's g-suit, except that, for the giraffe, the g-suit is activated all the time.

Imagine a blood vessel 17 feet long, extending from the ground vertically upwards. Assume (for simplicity's sake) that the blood isn't flowing. The first term in Bernoulli's equation, $v^2/2$, where v is for velocity, is thus zero. In the second term, gz, g is gravity and z is height, while in the third term, p/ρ, p is pressure and ρ (rho) is density. Bernoulli's equation tells us (within certain assumptions) that the sum of the three terms is everywhere equal. Now consider the conditions at the very top and very bottom of our 17-foot-long blood vessel. Since the first term is zero and the density of blood is constant, Bernoulli's equation tells us that the pressure at the bottom of the vessel has to be much greater than that at the top to make up for the gz (height) term. To contain that massive pressure, giraffes have evolved the built in g-suit described above. This also helps explain why the giraffe's heart is so thick. While the heart is much higher than the lower legs, it is still six feet or more below the head in an adult giraffe and thus subject to much higher hydrostatic pressures.

The above discussion doesn't take into account what happens when the giraffe begins moving its head about. Imagine the changes in pressure that have to be dealt with when, for example, the giraffe lowers its head 17 feet in order to drink from a stream and then when he raises it back after drinking. The giraffe has evolved some ingenious circulatory mechanisms to deal with these everyday realities—even though one of them isn't an extraordinarily large heart.

5. AC/DC

$$v(t) = v_{\text{peak}} \sin(\omega t)$$

Voltage of an Alternating Current

The voltage, $v(t)$, of an alternating electrical current varies sinusoidally with time, t, in this equation. The angular frequency is ω, and the peak voltage is v_{peak}. Alternating current that follows this equation is said to have a sinusoidal waveform. Other alternating current waveforms, such as triangular or square waveforms, are sometimes utilized, depending on the application.

Let's face it: the modern world runs on alternating current (AC). Household appliances from your vacuum cleaner to your refrigerator, toaster, and TV all utilize AC. The power line running into your house from the local electric utility delivers AC too, as do all the power lines strung or buried across your town. But the world runs on direct current (DC), too. Your car's electrical system is DC, and all of your portable electronic gizmos, from your iPad to your laptop computer, cell phone, and digital camera, run on DC as well. These days, we tend to use the system, AC or DC, that best fits the situation at hand, but it wasn't always that way. AC and DC were once locked in a bitter power struggle for dominance of the fledgling electric power industry. Let's travel back in time to the days of Thomas Edison, George Westinghouse, and the War of Currents.

Ask any American to name a famous inventor, and the likely response would be Thomas Edison, even though the Wizard of Menlo Park died long ago, in 1931. The inventor of the phonograph and the first practical electric light bulb, Edison held over a thousand U.S. patents. He also developed and deployed the first electrical power distribution system in the world. It used direct current. Edison invented his incandescent light bulb in 1878. To make electric light practical, he realized a system was needed to distribute electric power to homes and businesses. In 1882, he unveiled the world's first electric power system, which provided DC at 110 volts to 59 customers in Manhattan.

Another talented American inventor and businessman, George Westinghouse, was also a major player in developing electric power systems. Born in 1846, the year before Edison, Westinghouse had by 1869 invented a compressed air system for braking trains that allowed the train's engineer to brake all the cars simultaneously. This represented a huge advance in train safety, and the braking systems used on modern trains are descendants of Westinghouse's invention. By the time Edison produced his DC power distribution system in Manhattan, Westinghouse was already interested in electric power. He analyzed Edison's DC system and found it to be wanting. Edison's low-voltage DC was inefficient and was particularly poor at transmitting power over long distances. To provide power to a large community, Edison's system would have required power generation plants located every few city blocks.

In electric power transmission, the power, P, that is transmitted is equal to the current, I, times the voltage, V ($P = IV$). A waterfall analogy is sometimes used to explain current and voltage, where the amount of water flowing over the falls represents current, and the height of the falls represents voltage. If you double the height of the falls (voltage), you can cut the amount of water (current) in half and still get the same power.

Edison's DC transmission system operated at 110 volts, since that was what was needed to operate the light bulbs at his customers' homes and businesses. Transmitting sufficient power at 110 volts requires high current, and unfortunately, that results in severe losses: the amount of power lost in transmission, due to heating of the conducting wire, is proportional to the *square* of the current. If you double the current, you thus quadruple the power lost. This was well known at the time, but there was little that Edison could do about it. He would have liked to transmit power more efficiently at higher voltages and lower currents, but those higher voltages were incompatible with his customers' needs (electric lights), and Edison had no practical way to manipulate voltages in his DC system. The ability to change, or transform, the voltage in an electrical system simply and inexpensively is what gave AC its huge advantage.

In alternating current, the voltage changes with time; our equation at the beginning of this section shows one common form this can take. In this case, the voltage is sinusoidal, alternating between positive and negative. The first practical AC transformer was demonstrated in 1881. A transformer is a simple device that allows the voltage of an alternating current to be increased or decreased. In most modern electric power transmission systems, after power is generated (by burning coal, for example, and using the heat to spin a turbine connected to a generator), a transformer then steps the voltage up to a very high level—often thousands of volts—so that the power can be efficiently transmitted (at low current) over long distances. At the other end of the power line, another transformer steps the voltage back down to the levels required by the customer. In the United States, 120 volts is the standard. Most of the rest of the world uses 220 or 230 volts; this is why your hair dryer may catch fire if you try to use it in Europe without first stepping the voltage down through a transformer.

In 1886, Westinghouse constructed the first AC power distribution system, in Massachusetts. The system used transformers to manipulate the voltage as required. Westinghouse was able to overcome two technological challenges with AC systems: a means of metering or measuring the power delivered to a customer, and electric motors compatible with AC power. Westinghouse obtained the rights to the first AC motors, which had been patented by the brilliant Serbian inventor Nikola Tesla. Tesla had worked for Edison, but the two famously split after disputes over technology and over Tesla's paltry salary. Decades later, Edison said that the biggest mistake he ever made was in disregarding Tesla's work.

By the late 1880s, Westinghouse was rapidly expanding his AC systems, which competed directly with Edison's earlier DC scheme. These days, you can tell how old someone is by the technology wars they have lived through. Too young to remember the battle between eight-track and cassette tapes? How about Betamax versus VHS? Or Blu-Ray versus DVD? The War of Currents between Edison's DC and Westinghouse's AC was the granddaddy of them all. It set the bar quite high in terms of the sheer absurdity and viciousness of it all. Looking back, it's pretty clear that Edison must have realized that his system was inferior, but that didn't stop him from going to ridiculous lengths to discredit the AC approach, and even Westinghouse himself.

Edison's main argument against AC was that it was unsafe. The electric chair, which has largely been supplanted by lethal injection but remains the method of execution for condemned prisoners in several U.S. states, was actually developed as part of the War of Currents. Edison opposed capital punishment, but he did not let his personal beliefs get in the way of an outrageous scheme to discredit Westinghouse and AC. How better to highlight the dangerous nature of AC than by having states adopt it as their method of execution? While he succeeded in getting electrocution—utilizing AC—adopted for capital punishment, Edison failed in his attempt to popularize his pet name for the process. We don't refer to those subjected to the electric chair's lethal currents as having been "Westinghoused."

Strangely enough, the main technological drawback to direct current, the lack of a means for changing voltages, was eventually overcome. Voltage conversion technology has developed steadily over the years, and in 2014 high-voltage DC power transmission has advantages over conventional AC and is frequently chosen, especially for very long distances. Somewhere, Thomas Edison is smiling.

6. The Doppler Effect

$$f' = f_s \left(\frac{1}{1 - \frac{v_s}{v}} \right)$$

The Doppler Effect

The apparent frequency, f', of sound waves travelling from a source moving at speed v_s towards a stationary observer may be calculated from the actual frequency of the sound source, f_s, and the speed, v, at which the waves are travelling. If the source is moving away from the stationary observer, the minus sign in the above equation becomes a plus. When the source is stationary and the observer is moving, the equation has a somewhat different form.

The American Physical Society has a bright red bumper sticker that reads, "If this sticker is blue, then you are driving too fast." Physicists get the joke, which plays off of the famous Doppler effect.

We all know the Doppler effect—through our ears, however, and not our eyes. A fire truck, siren blaring, speeds down the road as you stand on the sidewalk. As the truck roars past you, the pitch of its siren drops noticeably—it sounds different going away from you than it did approaching you. That's the Doppler effect. But did you also notice a change in the fire truck's bright red color? Just as surely as the sound changes, the color of the truck changes too, in both cases due to the Doppler effect. We all notice the change in sound, but no one has eyes sharp enough to detect the exceedingly small color change involved in our fire truck example.

Frequency is at the heart of the Doppler effect, which was first described in 1842 by Austrian physicist Christian Johann Doppler (1803–53). He used the effect that now bears his name not to describe sound but to explain the color of the stars based on their velocities relative to an observer on the Earth. Sound waves change *pitch* based on the relative velocity of the sound source and the observer. Light waves, in analogous fashion, change *color*. Pitch and color are what we perceive as a function of the frequency of sound or light. If you drop a series of pebbles into a still pond, waves of water will radiate outward across the surface. The number of wave crests that pass a given point on the surface per unit time is the frequency of the waves. Now imagine that you are on a boat traveling towards the point at which the pebbles hit the water. The frequency with which you pass the waves will increase, and that increase is predicted by the Doppler effect. Likewise, if the boat is moving in the opposite direction, the frequency of waves passing it will decrease.

Sound and light consist of waves too. These waves have frequencies just like the waves on the surface of the pond. Sound is waves of vibrating air molecules. Our ears detect different frequencies, and our brains interpret them as differences in pitch. Light has a wavelike nature as well. Here, differences in frequency are perceived as different colors.

Doppler's work related to light waves—he was interested in the relative motions of stars and planets—but his results were quickly extended to sound waves by others. The Dutch chemist and meteorologist C. H. D. Buys-Ballot organized a famous demonstration in 1845 in which a trainload of trumpeters, all playing the same note, was transported past a group of stationary observers on a platform at the station. The pitch of the musical note heard by those observers changed, as predicted, as the train first approached and then moved away from the platform.

In the Doppler effect, the detected frequency is the frequency at which the observer meets the waves traveling outwards from the (moving) source. The detected frequency, f', can be calculated from the emitted frequency, f_s, of the source using our equation above. In our equation, v is the speed at which the waves are traveling (in our tuning fork example below, this is the speed of sound in air), and v_s is the speed at which the sound source is approaching the observer, as shown schematically in figure 3. The equation predicts the frequency, f', that the observer perceives.

Imagine that you are standing beside a road. Moving towards you at 70 miles per hour (31.3 meters per second) is a vehicle with an A440 tuning fork on its roof. The speed of sound in air is 343 meters per second. If you plug those numbers ($f_s = 440$ Hz, $v = 343$ m/s, and $v_s = 31.3$ m/s) into the equation, you will find that the detected frequency of the 440-Hz tone as the tuning fork approaches you at 70 miles per hour is about 484.2 Hz. As the car with its tuning fork passes and recedes into the distance, the Doppler effect predicts that the observed frequency will drop to 403.2 Hz.

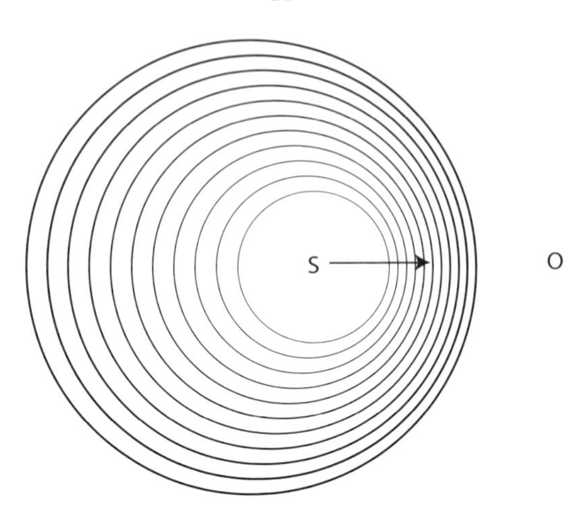

Figure 3. Schematic representation of the Doppler effect with a sound source (*S*) approaching a stationary observer (*O*). The apparent frequency of the sound waves (circles) reaching the observer is greater than their actual frequency.

The frequency shift in this example, from 440 Hz up to 484 Hz, is more than a musical semitone, or half tone. Frequency shifts much, much smaller than this are clearly detectable by the human ear. For example, the 49th key on a piano—the A above middle C—has a frequency of 440 Hz. The keys on either side of it have frequencies of 466 and 415 Hz.

The tuning fork in our example sounds different when it is approaching you than when it is moving away. Anyone with anything approaching normal hearing will notice the difference. But as noted earlier, Doppler's work also tells us that the *color* of the vehicle will shift as it speeds first towards us, then away from us. Our ears hear the difference, so why don't our eyes see it as well? It's because light waves travel so much faster than sound waves.* Light travels at about 300 million meters per second, which is roughly a million times as fast as sound travels in air. The ratio of v_s to v in Doppler's equation for the sound shift in the tuning fork example was about 10% (resulting in, as noted, a frequency shift of more than a musical semitone). But the ratio of v_s to v for the shift in the frequency of light waves is only about one part in 10 million. No one's eyes are that good. But plenty of human-made instruments *are* that sensitive.

*The Doppler effect as applied to light waves uses equations that are somewhat more complicated. Qualitatively, however, the effect is the same as in the sound wave examples in this story.

The Doppler effect has lots of applications in instruments, including everyone's favorite, the policeman's radar gun. The radar gun uses the Doppler effect not on sound or visible light waves but on microwaves. These have much higher frequencies (and velocities) than sound waves—a radar gun might operate at, say, 10 GHz or 10 billion cycles per second. Their operating principle is relatively simple. The radar gun, often a handheld unit, directs a beam of microwaves at the target. Those microwaves travel at the speed of light, about 300 million meters per second. But the car is only traveling at a tiny fraction of that speed, say 70 miles per hour, or 31.3 meters per second. In other words, the microwaves are traveling nearly 10 million times as fast as the car. So the microwaves bounce off the speeding car and are reflected back to the radar gun, which captures the reflected waves and measures their frequency. The change in frequency from the transmitted to the reflected microwaves caused by the Doppler effect is used to calculate the speed of the target. Because the speed of the car is so small compared to the speed of light, the frequency change is also tiny. Nonetheless, such frequency changes are routinely and accurately measured, and have been for many years: the first police use of this type of technology against speeding cars in the United States was in 1954. Christian Johann Doppler would have been proud.

7. Do I Look Fat in These Jeans?

$$\text{BMI} = \frac{\text{mass}}{\text{height}^2}$$

Formula for Body Mass Index

The body mass index (BMI) is the ratio of mass in kilograms to the square of the height in meters for an adult human being. To calculate BMI using U.S. customary units requires a units conversion. If the height is in inches and the weight is in pounds, the ratio in the above equation must be multiplied by 703.

Pornography, as Supreme Court justice Potter Stewart famously remarked, may be hard to define, "but I know it when I see it." The same thing goes with being overweight—we know *that* when we see it, too. But then again, unlike pornography, we can quantify being overweight. For that, we have the body mass index. All you need to know is how tall you are and how much you weigh. Your BMI will then tell you whether you are normal, underweight, overweight, or obese. Sometimes.

The formula for the BMI was published way back in 1870, by the Belgian statistician, astronomer, and mathematician Adolphe Quetelet (1796–1874). In our equation above, body mass is in kilograms and height is in meters; this is Quetelet's original formula. If you want to calculate your BMI in customary U.S. units, you will need to multiply by a constant, 703, to account for the different units. Thus, BMI = 703 × (weight in pounds) / (height in inches)^2.

Quetelet was looking for a single number with which to categorize individuals with respect to their body weight. Knowing only how much someone weighs tells us little about the appropriateness of that person's weight. Someone weighing 180 pounds could be underweight, normal, overweight, or even obese—depending on how tall that person is. Realizing this, Quetelet set about to find a formula that took height into account when evaluating weight. He found that dividing body mass by the square of the height did a reasonable job in compensating for differences in height—and thus allowed him to compare folks with the use of a single number. There was no basis in any sort of theory for his BMI formula, however; it just seemed to work reasonably well. It has been argued that the exponent 2 in the equation is probably too low: perhaps it should be something closer to 2.6. That we use 2 may be due at least as much to the convenience of a whole number as to anything else. In the days before electronic calculators, it was much easier to simply square a number than it was to raise it to something like the 2.6 power.

At any rate, what became Quetelet's index—now more familiar as the BMI—caught on. Today, insurance companies sometimes use the BMI to determine risk categories for health insurance. If your BMI places you in the "morbidly obese" category, for example, you may find yourself paying higher insurance premiums. Researchers also use BMI to study the relationship of weight to all kinds of things, such as income, education, or place of birth. The BMI is not without controversy, however, especially when it comes to classifying individuals as normal, overweight, or obese.

At six feet six inches tall, with a listed playing weight of 215 pounds, Michael Jordan's BMI of nearly 25 placed him squarely on the boundary between "normal" and "overweight." To suggest that the man many believe was the greatest basketball player of all time was overweight as a player is patently ridiculous. They didn't call him "His Airness" for nothing. Likewise, Serena Williams' BMI of about 26 lands her smack in the middle of "overweight." So one of the greatest tennis champions of all time, who covers the court with catlike quickness, is overweight? Hardly. And yet, another woman of Serena's height and weight (five feet nine inches and 175 pounds), with a more slender frame and a more sedentary lifestyle, might indeed be considered overweight.

It is clear that there is more to being overweight than just the BMI. Athletes tend to have more muscle mass than the rest of us, and thus higher BMIs, since muscle is quite dense. The elderly often become shorter with age. This causes their BMI to increase even if their body weight does not. Optimum body weight varies with age. BMI was originally developed for young adults through late middle age. Its use for children and the elderly requires modification.

Researchers who use the BMI for other purposes have repeatedly cautioned against its indiscriminate use for categorizing individuals without regard to their lifestyle, age, frame type, and other factors. In this respect at least, the BMI is not unlike various other metrics that have been developed to quantify complex human phenomena. The intelligence quotient, or IQ (discussed in chapter 35), a measurement similarly developed for narrow purposes, has been inappropriately applied in so many different situations it would be difficult to count them all, and so it may be with the BMI.

BMI is not the only way to quantify whether someone is overweight. Body fat metrics such as skin-fold measurements or underwater weighing are both useful. Unfortunately, these require more work. BMI is extremely easy to measure. Thus, it gets used all the time, including times when it shouldn't.

Most of us have heard stories on the news about how fat, as a people, we Americans have become. As likely as not, the statistics behind those stories are based on the BMI. For example, the percentage of overweight Americans (BMI greater than 25), aged 20–74, increased from 45% in 1960–62 to 65% in 1999–2002. Over the same 40-year period, the percentage of obese Americans (BMI greater than 30) increased from 13% to 31%. By this measure, nearly a third of American adults are obese. The percentage of overweight children, aged 6–11, increased from 4% in 1963–65 to 16% in 1999–2002.

The BMI may not be a perfect measure, but data like those in the above category are nothing if not a wake up call to the American people. First Lady Michelle Obama headed up the "Let's Move" campaign, designed to combat weight problems in children. That the First Lady chose this as her signature issue should be some indication of the seriousness of the problem. Only time will tell if she, or anyone else, can make a difference.

8. Zeros and Ones

$$10 + 10 = 100$$

An Example of Binary (Base 2) Arithmetic

In binary arithmetic, 10 is equal to 2 in the familiar base 10 system. Each digit in a binary number is either a 0 or a 1. Thus, $10 + 10 = 100$ in binary is the same as $2 + 2 = 4$ in base 10. Binary operations such as multiplication and division may also be performed. Base 10 fractions and decimals may be expressed in binary form.

"Answer the question, 'Yes' or 'No!'" barks the district attorney at the witness, imploring a black or white answer, with no room for equivocation. In the cold logic of the courtroom, there are no shades of gray. The defendant is either guilty or not guilty: "maybe" is not an option. All this would have been right up George Boole's and Claude Shannon's alley.

George Boole (1815–64) was an English mathematician and philosopher who codified a system that has come to be known as Boolean logic. When Boole died, the phenomenal utility inherent in his system remained undiscovered—it was simply a clever bit of mathematical wizardry set forth in some dusty old journals. And there it would remain for more than 70 years, until 1937, when Claude Shannon (1916–2001) wrote a master's thesis at MIT entitled "A Symbolic Analysis of Relay and Switching Signals." Shannon's work combined Boolean algebra with binary arithmetic and essentially ushered in the era of modern digital circuit design. The Harvard psychologist Howard Gardner has remarked that Shannon's work was "possibly the most important master's thesis of the century."

Black or white. Yes or no. True or false. On or off. Zero or one. All modern computing devices, from your laptop to your cell phone, your car, and even your toaster, operate on a logic-based on a system of zeros and ones. The slogan on perhaps the nerdiest T-shirt ever printed goes like this:

There are only 10 kinds of people in the world:
Those who understand binary and those who don't.

This is just binary math, played for laughs. For 10, in binary math, is equal to 2 in base 10 math.

In the preface to this book, we noted Lord Kelvin's tongue-in-cheek definition of a mathematician as someone to whom the solution to a complex equation is just as obvious as it is to a normal person that $2+2=4$. Well, our equation above, $10+10=100$, is just that: $2+2=4$, except that our version here is in binary arithmetic. In the binary system, each digit of a number is either a 0 or a 1.

Remember back to when you first learned how decimal numbers work. Let's say you were born in 1985. In that year number, the 5 tells you how many ones are in the number, the 8 tells you how many tens there are, the 9 tells you how many hundreds there are, and the 1 tells you how many thousands there are. Thus, 1985 is just $1 \times 1000 + 9 \times 100 + 8 \times 10 + 5 \times 1$. As you move from right to left in a decimal number, the amount being expressed by the next digit is always 10 times as great as the preceding digit.

Binary numbers work the same way, except that instead of 10 times, we have 2 times. The rightmost digit in a binary number tells you how many ones there are in the number—either 0 or 1. The next digit to the left tells you how many twos there are (0 or 1), the next digit is how many fours there are (0 or 1), the next is how many eights (0 or 1), and so on. Any decimal whole number can be expressed as a unique binary number. (Fractional numbers work in binary too, although it's a little more complicated.)

For example, the decimal number "23" is equal to $1 \times 16 + 0 \times 8 + 1 \times 4 + 1 \times 2 + 1 \times 1$. Its unique binary expression is therefore 10111. What could be simpler than that? Or at least that's how computers regard the situation. Computers can gobble up and manipulate long strings of zeros and ones at speeds that boggle the mind. Inside a computer, all numbers are expressed in binary form.

This is because a binary digit, 0 or 1, is just like an electronic switch, which is either on (1) or off (0). When Claude Shannon wrote his thesis at MIT in 1937, electronic switches were clunky mechanical devices. Since the advent of the integrated circuit in the late 1950s, electronic switches in the form of transistors can be made vanishingly small, with no moving parts, such that billions of them can be fabricated into a single silicon chip. (See our discussion of this in chapter 10.)

So it's simple enough to represent numbers, even impossibly large ones, inside a silicon chip. Numbers are just represented in binary as a series of switches, each of which is either on or off. Representing numbers in binary is just a small part of the story, however, and this is where George Boole and Claude Shannon come back in. Boole invented a system of logic for operations involving zeros and ones. Shannon, all those years later, realized that the rigor of Boole's system could be applied to electronic circuits. One of the original applications was to simplify telephone switching circuits. But Shannon also showed that electronic circuits could be used to solve Boolean logic problems.

Here's an example. Let's say a wealthy alumnus of a large university wants to offer a scholarship to a female student, provided that she is married, a veteran of the U.S. military, and majoring in either biology or chemistry. And so the university wants to make a list of all the eligible students. Each of the factors involved here can be stated in binary terms: Female? Yes or no. Married? Yes or no. Veteran? Yes or no. Biology major? Yes or no. Chemistry major? Yes or no. To be eligible for the scholarship, a given student's answers to each question must be yes, except for the last two questions involving the student's major. If the answer to either one of the last two questions is yes, that's good enough. A logic statement for this database search might look something like this: (Female) and (Married) and (Veteran) and (Biology or Chemistry). Here we employ the formal logical (Boolean) meanings for *and* (both must be true) and *or* (only one or the other need be true).

Computer database software is designed to perform logical searches such as this, and it wouldn't take more than a few seconds to find all the students at the school eligible for the scholarship. The lucky winner would have, in addition to her wealthy benefactor, both George Boole and Claude Shannon to thank.

9. Tsunami

$$y = a\sin\left(\frac{2\pi}{L}(x - ct)\right)$$

Sinusoidal Equation of a Wave as It Varies with Time, *t*

The vertical position of any given point on the wave is y, and x is the horizontal position of the same point. The constants in the equation are a, half the distance from the crest to the trough of the wave; L, the distance between successive crests; and c, the wave's speed.

It's a safe bet that lots of folks, back in 2004, had only the vaguest of ideas as to what constituted a tsunami or of how destructive one could be. That was before the devastation of the Indian Ocean tsunami of December 26, 2004, which killed an estimated 230,000 people, making it the deadliest tsunami in history and one of the 10 deadliest natural disasters of all time. More recently, Japan was ravaged by a massive earthquake just off its northeast coast in March 2011. The earthquake and resulting tsunami killed many thousands and also resulted in what was probably the second worst incident in the history of the nuclear power industry (excepting only Chernobyl), when some of the reactors at Fukushima partially melted down in the wake of the earthquake and tsunami.

All of that, for a simple ocean wave that might not be much more than one foot high out in the middle of the ocean. So then why do tsunamis have the potential to do some much damage?

According to the *American Heritage Dictionary*, a wave is a "disturbance travelling through a medium in which energy is transferred from one particle in the medium to another without causing any permanent displacement of the medium itself." Waves can travel through all sorts of different media, but here we're concerned with the waves that travel through water. The definition above refers to energy being transferred without "permanent displacement of the medium itself." Have you ever been out on a boat when a wave passed by? The wave gives the illusion that the water it contains—the water in the wave—is travelling horizontally across the surface of the water, but that isn't true, and the boat proves it. The boat bobs up and down as each wave passes, but it moves very little in the horizontal direction—the direction the wave is travelling.

A wave moving across the water is a little bit like a row of dominoes. When you knock over the first domino, it knocks over the second one, which topples the third, and so on down the row. Each domino, in falling, transfers its energy to the one next to it, just as each particle of water in a wave transfers its energy to the particle of water next to it. And as with the dominos, each individual particle of water moves very little in the direction of the wave.

As a wave crosses a body of water, it takes a sinusoidal form. In our equation above, y is the vertical position of a point on a wave, x its horizontal position, and t is the time. The other parameters, a, L, and c, are what make one particular wave different from another. The parameter L is the wavelength (or distance between the crests of successive waves), while c is the wave's speed (the rate at which it moves across the surface of the water). The parameter a equals half the vertical distance from the crest of a wave to its trough. Thus, a wave in water extends upwards to its crest by $+a$ from the undisturbed surface of the water, and downwards to its trough by $-a$ from the undisturbed surface.

Waves on the ocean or on large lakes are normally caused by the wind. The distance between successive crests of such waves is typically about 100 feet or so; this is the wavelength. Tsunamis, in contrast, have truly enormous wavelengths of 50 to 100 miles. Even though the height of tsunami waves out in the middle of the ocean might be only a foot or two, with a wavelength of 50 miles or more they contain a staggering amount of water. Tsunami waves also move very fast. Wind-induced waves generally move at 30 miles per hour or less. Tsunami waves can travel at speeds of up to 500 miles per hour—roughly the cruising speed of many jet airliners.

The massive volume of water in a tsunami, moving very fast, contains an astonishing amount of energy. That energy gets dissipated on whatever is in the way when the waves hit the shore. We've all seen the videos from Japan and the resulting devastation.

Something strange happens to water waves when they approach a coastline, as anyone who's ever visited the ocean knows. The waves slow down, and they get taller. Wind-generated waves that might be only a few feet tall in the deep ocean can become dozens of feet high as they approach the shore. A lot depends on the characteristics of the shoreline in question. As any surfer could tell you, some beaches are better than others at generating those tall breaking waves that surfers live for. Because of their enormous wavelengths, and the high speeds at which they travel, tsunamis pose a particular threat as they approach land.

Most tsunamis are caused by earthquakes beneath the ocean's floor. The earthquake that caused the 2004 Indian Ocean tsunami caused a section of seabed about a thousand miles long to shift sideways 30 or more feet and upwards by more than a dozen feet. This sudden motion of the ocean floor lifted an enormous column of water upwards. The result was a tsunami. That huge column of water pushes upwards and then moves outwards, away from its subsea source.

The slight bulging upward of the water at the surface is really just the tip of what might be thought of as an enormous liquid iceberg. As the wave approaches the shore, that column of water starts to get taller and to slow down as the ocean gets shallower and shallower. Double trouble. As the front end of a tsunami wave slows down, gets taller, and crashes ashore, the rest of the wave, still moving fast, rushes in behind. Remember, tsunamis can have wavelengths of 50 miles or more. This is one reason why tsunami surges last for many minutes.

Because there are so many variables, it is difficult to predict exactly how a tsunami will behave once it reaches the shore. Detecting tsunami waves in the deep ocean, however, and predicting where they will make landfall, is relatively simple. Such waves produce small but easily measureable pressure spikes as they make their way across the ocean. Pressure sensors deployed on the ocean floor can detect these spikes and wirelessly relay warning signals to those regions in the path of the wave. As more and more tsunami detection and warning systems are installed, and as people are educated about what to do and where to go when they hear a tsunami warning, the potential to avoid widespread loss of life increases. Protecting the shoreline itself is a much more difficult task.

10. When the Chips Are Down

$$P_n = P_0(2^n)$$

An Equation for Exponential Growth

In this example (one equation among many), P_0 is the current value of something that (in this case) doubles with each passing unit of time. If the unit of time is, for example, years, then P_n is the value of P after n years. In general, exponential growth occurs when the growth rate of something is proportional to the current value of that thing. In our example here, since P is doubling every year, its growth rate is 100% per year.

The horse revolutionized agriculture, and the steam engine powered the industrial revolution. In like fashion, the age of electronics was ushered in by the solid-state transistor. The first solid-state transistor is generally agreed to have been produced in the United States at AT&T Bell Labs in 1947. Individually packaged solid-state transistors hit the market soon thereafter, but the real revolution in electronic devices began after the technology to integrate multiple transistors (and other electronic devices) on a single piece of material, a silicon wafer, was developed in 1958.

Almost immediately, a race was on—one that continues to this day—to miniaturize these devices. As the individual elements in integrated circuits got smaller and smaller, more and more of them could be packed into a silicon chip of a given size. The result, as everyone knows, is that electronic devices continue to shrink in size even as they expand in capability.

A transistor is an electronic device used to either amplify or switch an electronic signal. Along with other basic electronic devices such as resistors, capacitors, and diodes, transistors are the building blocks of modern electronic devices—and thus of modern life itself. To say that transistors are ubiquitous is to state the obvious. It has been estimated that about one billion transistors were produced in 2010 for each human being on our planet. How's that for ubiquity?

In April 1965, not long after the birth of the integrated circuit, Gordon Moore, the cofounder (in 1968) of chip maker Intel, wrote a paper in *Electronics Magazine* in which he stated that the rate at which chip-making technology was improving seemed to be following a quantifiable trend. Moore noted that the "complexity for minimum component costs" had increased (since the birth of the integrated circuit, about seven years earlier) at "a rate of roughly a factor of two per year." He contended that this trend was likely to continue "for at least ten years." This allowed Moore to put forth the following: "That means by 1975, the number of components per integrated circuit for minimum cost will be 65,000. I believe that such a large circuit can be built on a single wafer."

This was a bold prediction. It is likely that a lot of folks thought that Moore was either nuts, or shamelessly shilling for his industry (he was, after all, a chipmaker himself), or both. What Moore was predicting was more than a thousand-fold improvement in performance over a 10-year period. In 1965, the most complex integrated circuits that Moore and his colleagues could produce contained about 30 elements. Moore himself has noted that in 1965, it was cheaper to cobble together your own (nonintegrated) circuit from individual components than it was to pay for an integrated circuit.

Our equation here is one way to mathematically render what came to be known as Moore's law. (The term "Moore's law" was coined a few years after Moore's original article by Moore's friend, the Cal Tech professor Carver Mead.) If P_0 is the current performance of something (say, the number of electronic elements that can be squeezed into one square centimeter on a chip), the equation predicts that the future performance n years later, P_n, rises exponentially. The future performance, in this case, equals the current performance times two raised to the nth power.

The exponent n in the equation indicates that the performance doubles every year. Others have stated that the performance of electronic devices is doubling more slowly than that, perhaps every 18 months—and thus, perhaps the exponent should be a little smaller. In 1975, Moore revised his 1965 prediction to a doubling of performance every two years. In this case, our equation above would become

$$P_n = P_0 \cdot 2^{n/2}.$$

In any event, the predictions of Moore's law have been eerily accurate. As each succeeding generation of computer chips has been shown to roughly follow Moore's original prediction, the prediction itself has been extended into the future. In 2008, Intel suggested that Moore's law would continue to hold until 2029. Some prognosticators have been more optimistic, others less so. The technology currently used to manufacture computer chips will eventually run into fundamental limits as transistors approach the size of individual atoms. Other technologies, however, may mature in time to allow the improvements predicted by Moore's law to extend far into the future.

Moore's law is regarded with something approaching religious fervor in the electronics industry. In 2005, Intel offered and paid $10,000 to the first person that was able to provide an original copy of the 1965 edition of *Electronics Magazine* in which the famous prediction was made. What was in the beginning merely a simple observation has to some extent been turned on its head: Moore's law says we should be at such-and-such level of performance by next year. By golly, we'd better get there! And it's not as if Gordon Moore had been the only person predicting, way back in the early days, phenomenal improvements in the performance of electronic devices. Others, such as computer pioneer Alan Turing, had made similar prognostications—in Turing's case, as far back as 1950.

Moore's law has become a stand-in for predicting the future performance of all kinds of things, especially things that tend to change very rapidly, and even Moore himself seems a little embarrassed by all the attention. "Moore's Law has been the name given to everything that changes exponentially," he sarcastically remarked. "I say, if Gore invented the Internet, then I invented the exponential."

11. A Stretch of the Imagination

$$F = -kx$$

Hooke's Law

The extension, or change in length, of a spring is proportional to the restoring force, F, exerted by the spring. The constant k is called the spring constant. The minus sign in the equation is necessary because the restoring force acts in a direction opposite to the direction in which the spring is changing length.

Robert Hooke is among the most enigmatic of the great scientists the world has known. Although his name is barely recognized today, except among scientists and engineers, he has been called "England's Leonardo" for his brilliance and versatility. His discoveries, such as the mathematical relationship stated in Hooke's law, changed the world and remain highly relevant more than 250 years after his death. And yet we can't even be sure what Robert Hooke looked like. Not a single portrait or drawing remains that is known with certainty to be his likeness. Although he died a rich man, the exact location of his grave is unknown.

When Hooke published what is now known as Hooke's law, he did so in Latin and in the form of an anagram: ceiiinosssttuv. Three years later, he provided the translation: *Ut tensio, sic vis.* It was common in those days for prominent scientists to publish findings in anagram form. In this way, they were able to establish the preeminence of their discoveries without revealing the details. *Ut tensio, sic vis*: As the extension, so the force. In other words . . .

When you pull on something, it gets longer. Everybody knows that. Whether it's a rubber band, a piece of rope, a steel I-beam, or a block of granite, if you pull on it, it will stretch. A lot in the case of the rubber band, very little for the steel I-beam. Humans have surely known this for a very long time, but it fell to the brilliant Englishman Robert Hooke (1635–1703) to quantify the relationship between the force applied to something and how much it stretches.

Hooke's law is among the most important relationships in the field of solid mechanics, or more precisely in the branch of solid mechanics known as elasticity. Hooke's law says that the extension, x (or change in length, or amount of stretch), of a piece of solid material is proportional to the force, F, that the material exerts in resisting the extension. The constant k is often called the "spring constant," and indeed, springs are perhaps the most common and easiest to understand of the many, many applications of Hooke's law. Consider an old-fashioned fish scale (not the modern electronic kind). This type of scale uses a spring and Hooke's law to keep fishermen honest. The heavier the fish that is hanging from the spring, the greater the force, F, exerted by the spring, and the more the spring stretches (x). The spring is calibrated—the calibrations are the numbers on the scale to indicate pounds or kilograms—through knowledge of the spring constant k. Thus, if a 10-pound fish stretches the spring inside the scale by an inch, the spring constant k is 10 pounds per inch.

Materials that obey Hooke's law (that is, most solids) are said to be "Hookean." Hookean behavior is also called linear-elastic, which just means that the relationship between force and change in length for such materials is linear: it follows Hooke's law. Hooke's law thus has tremendous practical applications; it describes how everyday objects such as cars, airplanes, bridges, and buildings change shape when forces are applied to them. Most metals, such as steel and aluminum, are Hookean, as are brittle materials like glass. Rubber, especially, and most plastics are non-Hookean. When you stretch a rubber band, its response is nonlinear. A graph of a rubber band's change in length versus the force it exerts is not a straight line.

When applied to materials, such as a bar of steel (as opposed to a coiled spring), Hooke's law is often written as

$$\sigma = E\varepsilon.$$

This equation is analogous to $F = -kx$. In this case, σ is the stress acting on the bar of steel (where the stress, σ, is the negative of the force, F, per unit area—the sign changes since σ acts on the solid, rather than vice versa), and ε is the strain, which is the change in length (displacement) x as a percentage of the original length. The constant E, called the elastic modulus, is analogous to the spring constant k. The elastic modulus of steel, about 30 million pounds per square inch (or 205 GPa in the metric system), is a number quite familiar to most mechanical and civil engineers.

Some materials behave in a more complex way than steel or glass. Such materials have more than one elastic modulus value. Wood is a great example. It is much stiffer (it has a larger E) with the grain than across the grain, as any carpenter could tell you. The elastic behavior of these materials can be explained with a more complex, multidimensional version of Hooke's law.

As befitting someone who's been compared to Leonardo, Robert Hooke moved with seeming effortlessness among a wide variety of intellectual pursuits. The breadth of his discoveries and contributions is breathtaking. He was among the first to create a practical, portable timepiece or watch. He was a prolific and important architect. He was a gifted instrument maker. Using his own microscopes, he studied and drew likenesses of a wide variety of plants and animals. Such studies led Hooke to coin the term "cell" for the basic structural unit of all living things. He also made contributions to the fields of paleontology, astronomy, and gravitation. With respect to gravitation, he appears to have come tantalizingly close to some of the discoveries that made Isaac Newton famous.

Late in life Hooke had a number of intellectual and personal disputes with Newton and others. There is at least some evidence that Hooke's relative obscurity today may have its roots in some of those disagreements. Recent scholarship, however, has somewhat restored Hooke's reputation. Even so, referring to the enterprising Englishman as "England's Leonardo" may be stretching things just a bit.

12. Woodstock Nation

$$P = \sum_{i=1}^{N} \rho_i A_i$$

Equation for Estimating Crowd Size

This equation can be used to estimate the size of a crowd from, for example, an aerial photograph. The number of persons in the crowd, P, can be calculated by dividing the area in question into i individual sections of area A_i. The population density of a given section is ρ_i (so many persons per unit area). Multiplying the population density times the area, and then summing over all of the i sections into which the crowd has been divided gives the total population of the crowd.

There are roughly seven billion human beings on Earth. That works out to about 122 persons for every square mile of land on our planet. But that's just an average, and it includes lots of land, such as Antarctica, where hardly anyone lives. Bangladesh is one of the world's most densely populated nations, at about 2,500 inhabitants per square mile. In the United States, we've got plenty of wide-open spaces to go along with our big cities, and on average, there are only about 83 Americans per square mile.

From time to time, however, and not unlike the citizens of other countries, large numbers of Americans like to pack themselves into relatively small spaces, such as sports arenas or concert halls. It's a fairly simple matter figuring out how many people have attended an event involving tickets and turnstiles. But what about other public events, such as political rallies, free concerts, and demonstrations? Estimating the attendance at such gatherings is far more difficult.

Is this a problem? You bet. The backers of political Candidate X would like you to believe that a huge throng of, say, 250,000 people attended their candidate's latest rally. Pshaw, say Candidate X's opponents: there couldn't have been more than 50,000 folks there, and most of them were probably bored out of their minds.

On August 28, 2010, Fox News commentator Glenn Beck held his "Restoring Honor" rally at the Lincoln Memorial in Washington, DC. Rally organizers had a permit for 300,000 people. Their estimate of 500,000 attendees far outstripped that number. CBS News put the attendance figure much lower, at no more than 96,000—or less than a fifth of the rally organizers' estimate. But CBS News is a competitor of Fox News.

Even when you take into consideration the biases of the crowd estimators in this case, how could they vary by a factor of 5? It turns out that estimating the size of a crowd like this is far from simple, even for an impartial observer.

The Tournament of Roses Parade has been held in Pasadena, California, on New Year's Day every year since 1890. Sometime around 1930, parade organizers began estimating the size of the crowd lining the 5½-mile parade route. These attendance estimates over the years have consistently ranged from 1 to 1.5 million. And not everyone believes them. In the 1980s, the organizer of a rival parade estimated the crowd at the Tournament of Roses Parade at only 360,000.

So what is the best way figure out how many folks attended such an event? Our equation provides an answer. One way to estimate P, the population of a crowd, is to divide the area of the crowd into i individual squares of real estate. If you can then estimate the population density, ρ_i, inside each of the squares of area A_i, it is a simple matter to sum them all up as shown in our equation above to get the total attendance.

For example, close to the stage at a political rally, the crowd might be packed in pretty tightly, at a density of about one person per 5 square feet. This means each person would occupy a square of real estate about 26 inches on a side. Further back from the stage, the density of the crowd might drop off to one person per 10 square feet (each person on a square about 38 inches per side). Move back a little more, and the crowd's density is even smaller.

By the way, the maximum crowd density, according to crowd experts, is about one person per 2.5 square feet, which puts each body on a square about 19 inches on a side. This is the sort of thing you might have to deal with right in front of the stage at a really popular outdoor rock concert. In case you've never been to such an event, you could get some idea how crowded that is by packing 90 people into a 15- by 15-foot room—a typical size for a secondary bedroom in an American home.

If you want to do a reasonable job estimating the size of a crowd, a good place to start is with aerial photographs of the event in question. On your photographs lay down a grid of, say, 50-foot squares (2,500 square feet per grid square). Then estimate the population density in each of those squares. Finally, sum up all the results for the individual squares to get the grand total.

The math is pretty simple, but the whole process is fraught with peril. Accurately estimating the density within each grid square is quite difficult. And any errors made there are magnified when you multiply the density by the number of grid squares. There are other problems, too. For example, since people are always moving around, and they can and do come and go from these events continually, who's to say your estimate corresponds to the maximum crowd?

The bottom line is, it's close to impossible to get really good figures for crowd sizes for parades, rallies, and the like. Crowd experts will tell you this, and most of the rest of us probably realize it deep down as well. Yet we continue to demand numerical crowd estimates from journalists, the police, and others. It's all part of our desire to quantify things. Disasters such as tornados and earthquakes are quantified in terms of the number of victims, and wars by the number of casualties.

And so it is with crowds. "By the time we got to Woodstock, we were half a million strong," sang Joni Mitchell. She didn't sing, "By the time we got to Woodstock, we were a really, really big crowd, although I'm not going to try and put a number on it because it probably wouldn't be accurate."

13. What Is π?

$$\pi = C/d$$

Equation for the Value of Pi

The most common way to define the irrational number π is as the ratio of the circumference of a circle, C, to its diameter, d. There are other ways to define π—for example, using trigonometric functions.

Cosine! Secant! Tangent! Sine! Three point one four one five nine! This cheer, which has propelled many a team to victory at the Massachusetts Institute of Technology, sets aflutter the hearts of geeks everywhere. Just thinking about it makes us want to go out and score one for the home team. But first things first.

It is hard to separate the ratio of the circumference to the diameter of a circle from the value of π (pi). An essentially equivalent equation to $\pi = C/d$ says that $A = \pi r^2$, where A is the area of the circle and r is the radius, or one-half the diameter. The constant π is arguably the most important and undoubtedly the most famous of all mathematical constants. A mathematical constant is just an unchanging value that shows up in a variety of different situations. In Einstein's famous equation $E = mc^2$, c is a mathematical constant—the speed of light.

Our equation $\pi = C/d$ provides a simple way to create π yourself. Take any round object, such as a bicycle wheel, a coin, or a can of tomatoes. Let's use the can of tomatoes for our example. Mark any point on the top circular edge of the can with a felt tip marker. Set the can on its side on a piece of paper so that the mark you made on the can is pointing straight down (at six o'clock) onto the paper. Mark that spot on the paper. Now roll the can across the paper until the mark on the can has made one full revolution. Mark the new spot on the paper. The distance between the two spots on the paper is just π times the diameter of the can. If you used a really small can, with a diameter of one inch, the distance between the two marks on your paper would be π inches.

So how many inches is that? Well, recalling our stirring cheer, that would be 3.14159 inches. But not exactly. It turns out that π is an "irrational" number. Mathematically, an irrational number is one that cannot be expressed exactly as the ratio of any two integers (whole numbers). For example, in the good old days before electronic calculators, π was often approximated as $3\frac{1}{7}$, which is the same as $\frac{22}{7}$. The difference between π and $\frac{22}{7}$ is very small—only about 0.00126. But is this difference really very small? It all depends on your perspective. In some applications, the difference might be negligible, but in others it could be way too large.

Who discovered π? No one really knows. Evidence exists that the ancient Babylonians, Greeks, Indians, and Egyptians all were aware that the ratio of the circumference to the diameter of any circle was the same, and that that ratio was a little bit larger than three. The earliest known recorded estimates of π date from about 1900 BCE. The Babylonians' estimate was $\frac{25}{8}$, and the Egyptians', $\frac{256}{81}$. Both estimates are within 1% of π.

Archimedes is on many short lists of the greatest mathematicians of all time. He is perhaps best known for discovering the principle of buoyancy (to wit: the famous "Eureka" story described in chapter 21), which discovery lies in the realm of physics, not mathematics. Nevertheless, Archimedes' accomplishments in pure mathematics were prodigious, and he was perhaps the first person to make a rigorous study of the value of π. He used a geometric technique, dubbed centuries later the "method of exhaustion," to painstakingly grind out ever closer estimates for π. The method works like this. Start with a circle. Draw a square surrounding the circle, such that the circle just barely fits inside the square. Now, draw a smaller square *inside* the circle, such that this square is just touching the circle at its corners, as shown in figure 4.

The area of the circle is obviously greater than the smaller square but less than the larger square. So, we have an upper bound and a lower bound on the area of the circle, and thus on π, since the area of the circle equals πr^2. The problem with this estimate is that the upper and lower bounds—the areas of the two squares—are too far apart to give a decent estimate. The outside square has an area equal to d^2, whereas the inside square is only half that, or $d^2/2$, where d is the diameter of the circle.

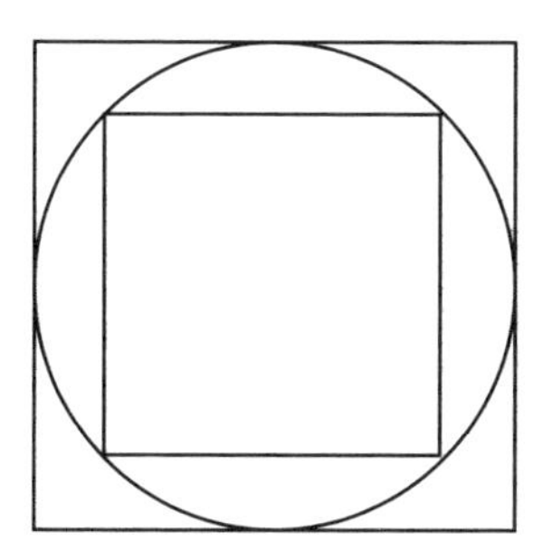

Figure 4. The method of exhaustion using squares

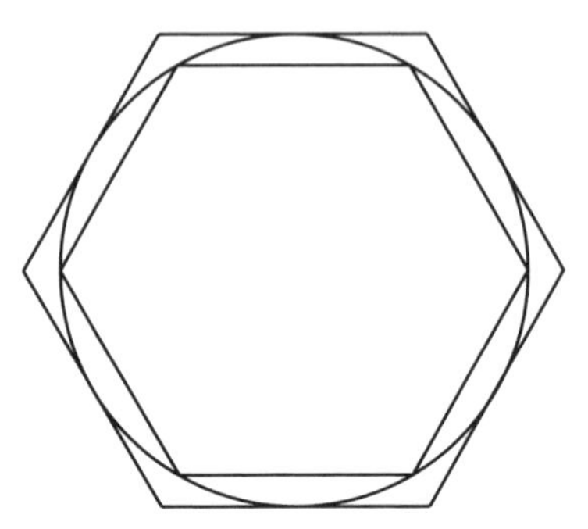

Figure 5. The method of exhaustion using hexagons

How can we make our estimate closer? Instead of a square (a four-sided figure), let's try a hexagon (six-sided). As with the square, draw a hexagon outside and inside the circle, as shown in figure 5. Here, the inside hexagon has an area of about $0.59d^2$, while the outside is about $0.65d^2$. Now we're getting somewhere.

Archimedes realized that the more sides the figures inside and outside the circle have, the closer together their areas are. Since the area of the circle is forever trapped between the areas of the two figures, Archimedes knew that the accuracy of his estimate of π was limited only by how large he was willing to make *n*, where *n* is the number of sides of the figure (square, pentagon, etc.). It is not for nothing that this technique is called the "method of exhaustion"! But Archimedes was as determined as he was intelligent, and he eventually showed, for $n = 96$, that π is greater than $3\frac{10}{71}$ but less than $3\frac{1}{7}$. The first of these numbers is about 0.02% smaller than π, while the second is about 0.04% larger. Not bad.

A long list of books, nonfiction and fiction, have been written about π. It has also inspired movies and music. And there are other cultural phenomena related to π. The record as of January 2010 for calculating π is almost 2.7 trillion digits. No one needs anywhere near this sort of accuracy for any sort of conceivable calculation. So why bother? Here's one reason: calculating π to a record number of digits requires a combination of fast computer hardware and efficient software—and therein lies the challenge, along with the bragging rights to the record holder.

Another bizarre π-related competition involves only the flesh-and-blood computers between our ears. The record for memorizing π, and then reciting it, has increased steadily over the years and now stands at 67,890 digits (as recognized by Guinness World Records). The Chinese record holder required just over 24 hours to recite all 67,890 digits without error, or one digit every 1.28 seconds. As Dave Barry would say, we're not making this up.

14. No Sweat

$$\frac{dQ}{dt} = hA_s(T_s - T_\infty)$$

Newton's Law of Cooling

The rate at which a body gains (or loses) heat is proportional to the difference between the temperature of the body and that of its surroundings. The time rate of heat flow into or out of the body is dQ/dt. The temperature difference between the body and its surroundings is $T_s - T_\infty$. The surface area of the body is A_s. The quantity h is the convection heat transfer coefficient, whose value depends on a number of factors.

In a town like Chicago, temperature swings of 130°F (−25°F in winter to +105°F in summer) over the course of a year are common. And yet if you are healthy, your body, through a collection of sophisticated mechanisms, maintains a core temperature of 98.6°F, regardless of how hot or cold it is where you are.

For the human body to maintain a constant body temperature, especially in a much colder or hotter environment, requires a constant battle against Isaac Newton's law of cooling. Newton's law of cooling is quite simple. It states that the rate at which an object gains or loses heat over time, dQ/dt, is proportional to the difference in temperature between the object and its surroundings, $T_s - T_\infty$. The larger the difference in temperature, the faster the object will tend to lose (or gain) heat. Set a glass of cold water and a cup of hot coffee on the kitchen counter. The cold water, at 40°F, is 38°F cooler than the 78°F air. The coffee, at 180°F, is 102°F hotter than the air. Come back in a couple hours, however, and both the water and coffee will be at room temperature, in accordance with Newton's law of cooling.

Your body, at a core temperature of 98.6°F, is a little more than 20°F warmer than the air in the above example. That you do not suffer a similar fate to that of your cup of coffee—that is, that you do not slowly cool to room temperature—is one of life's little marvels.

Newton's law of cooling tells us that the rate at which an object gains or loses heat is proportional to the difference in temperature between the object and its environment, but it doesn't tell us much about *how* those temperature changes take place. Heat flows in and out of the human body by at least four different mechanisms. Perspiration is perhaps the most familiar. (The others are radiation, conduction, and convection, stories we'll save for another time.)

Perspiration is a rather ingenious cooling mechanism. Few species take advantage of it to the extent that humans do. Horses perspire profusely, but the ability to perspire in other mammals such as dogs, cats, and pigs is quite limited. "Sweating like a pig" is thus a false simile. Perspiration cools us down because of a phenomenon known as evaporative cooling. When sweat, which is mostly water, forms on your skin and then evaporates, the water absorbs a lot of energy, which it carries away from your body as it changes from a liquid to a gas, thus cooling you down.

Imagine that you are going for a swim in a pool in Phoenix, Arizona, in July. The water in the pool is at 85°F, while the air temperature at the outdoor pool is 105°F. The air is thus much hotter than the water, yet as soon as you step out of the pool, your skin feels cool. This is the evaporative cooling effect. The water on your skin from the pool will dry very quickly in a hot, dry place like Phoenix, where the humidity is generally very low. So now your skin is dry, and you're standing outside in the sun on a 105°F day. Time for perspiration to take over.

Energy, mostly in the form of radiation from the sun, is entering your body at a significant rate, perhaps as high as several hundred watts. Your body has to get rid of that energy, or you will overheat—an extremely dangerous, potentially fatal condition. Fortunately, perspiration in humans is quite efficient. At about 93.2°F, your skin is usually about 5.4°F cooler than your body's normal core temperature of 98.6°F. Perspiration normally begins when the skin has been warmed, by the sun or something else, to about 98.6°F.

The amount of heat your body can get rid of through perspiration depends on lots of factors—things like how big you are and what you are wearing—but also very importantly on the humidity in the air around you. The lower the humidity, the more readily your perspiration evaporates and thus cools you down. When it's really muggy out, your sweat can't evaporate as well, and it just ends up soaking your clothes.

In any event, a person's body can easily cool itself at rates of 200 watts or more due to perspiration. We normally think of the watt as a measure for things like light bulbs, hair dryers, or electricity generators. A 100-watt incandescent light bulb generates mostly heat, as anyone who has ever touched one that has been illuminated for more than a few minutes is painfully aware. Only a few percent of the energy the bulb consumes is converted into light; the rest is wasted as heat. For the sake of simplicity, let's assume that a 100-watt light bulb generates 100 watts of heat. It is routine, then, for a human body to reject the same amount of heat, through perspiration alone, as is generated by two 100-watt light bulbs.

That's a lot of heat, and perspiration is thus, quite literally, a lifesaver. Think of it this way: if you suddenly lost the ability to perspire, then all of the heat from those two 100-watt light bulbs—heat your perspiration had been carrying away from your body—would no longer be able to escape, and your body would just keep getting hotter and hotter. In just one hour's time, 200 watts will increase the temperature of 150 pounds of water (about 18 gallons) by more than 4.5°F. The human body is mostly water, and so it is easy to see that losing the ability to perspire—losing the ability to get rid of those 200 watts—could kill you. A 4.5°F increase in body temperature in one hour, if unchecked, quickly becomes life threatening.

Most of us, without even realizing it, sweat out on the order of 600 milliliters per day (about 20 ounces), even on a day spent in a climate-controlled environment and without exercising. When you're exercising intensely in a hot environment, perspiration rates of 1.5 liters per hour (more than 1.5 quarts) can occur. If you don't replace all that lost liquid by drinking regularly during exercise, the result can be heat exhaustion or, in more extreme cases, heat stroke. The latter can be life threatening. It's important to get lots of exercise, but don't forget to hydrate. Your life could depend on it.

15. Road Range

$$P = \frac{E}{t}$$

Formula for Calculating Power

Power, P, is the rate at which energy, E, is consumed per unit of time, t. In SI units, energy is measured in joules, power in watts (joules per second), and time in seconds. In this equation work can be substituted for energy; work and energy have the same units (joules). Power is thus the rate at which work is performed or the rate at which energy is consumed. But it is impossible to convert all of the energy (in a gallon of gasoline, for example) into work. Work and energy thus differ by an efficiency factor that is often quite large.

In the early years of the 1900s, it was pretty clear that the automobile was going places and that it would someday dominate our transportation system. What wasn't so clear was which of several competing systems would win the battle to power the cars of the future. Would we all be motoring around on steam power, gasoline power, or battery power? Looking back, it's hard to imagine that this was much of a competition, but it was. Steam, gasoline, and battery power each had its pros and cons. Gasoline (and diesel fuel) won, in part because they were (and are) so inexpensive and plentiful, and because of the phenomenal amount of chemical energy stored per gallon of these hydrocarbon liquids. Much of that energy can be converted, in an internal combustion engine, into mechanical work, thus powering you down the road.

The steam-powered automobile is a quaint museum relic, but the battery-powered car never really went away. Battery-powered vehicles have always been successful in niche market applications such as golf carts, fork trucks, wheel chairs, and scooters. And battery-powered road vehicles have, from time to time, emerged from the shadows to make a serious run at the supremacy of the gasoline-powered car. By 2012, the battery-electric vehicle was in the middle of another such renaissance. Battery-gasoline hybrids such as the

Toyota Prius can be found on every corner, and battery-only vehicles such as the Tesla Roadster and Model S and the Nissan Leaf are already on the market, soon to be joined by a raft of competitors.

The on-again, off-again quest for a viable battery-powered car is inexorably linked to the development of a suitable battery to power the thing. Here's just one example of what automakers are up against. The much-hyped Chevy Volt, which hit the market in late 2010, is essentially a battery-electric vehicle with a small gasoline engine that can be used when needed to charge the batteries on the road. General Motors spent billions of dollars developing the lithium-ion battery pack on the Volt. The Volt's batteries weigh about 400 pounds, and GM claims they can propel the vehicle about 40 miles before it is necessary to fire up the gasoline engine and recharge the batteries. Forty miles is about the same distance that a regular (nonhybrid, non-electric) Honda Civic will travel on the highway, while consuming about *6 pounds* (one gallon) of gasoline.

So that is where we find ourselves, more than 200 years after the Italian physicist Allessandro Volta (1745–1827) first invented the electric battery, in about 1800. Roughly 100 years later, Thomas Edison was bemoaning the fact that batteries rarely perform as advertised, turning otherwise good people into liars. And 100 years after that, despite phenomenal improvements, we are left with the situation exemplified by the Chevy Volt.

Range. It has always been the Achilles heel of the electric car. The range of a gasoline-powered car is essentially unlimited. If your car has a 20-gallon tank, and you get 25 miles per gallon, you can drive 500 miles before refueling. Refueling involves spending a couple of minutes at a self-service gasoline station, which can be found within a few miles of just about any location in the United States and throughout the developed world. Once refueled, you're good to go for another 500 miles.

Let's compare that with the Tesla Roadster. This is a two-seat sports car, a gorgeous battery-electric machine with spectacular performance (0 to 60 in under four seconds) and a price tag of about $125,000. The 900-pound battery pack on the Roadster provides, according to Tesla Motors, a 245-mile range. Others disagree. The Tesla is a high-performance sports car, and if you drive it like one, you will get nowhere near 245 miles on a single battery charge. Some test drivers, having driven the car like the flat-out performance vehicle that it is, have reported range numbers as low as 100 miles, or lower. If you want to go 245 miles on a charge, you'd better drive like Grandma going to church on Sunday.

Would you buy a $125,000 car if the salesperson told you that you could go no more than 245 miles before refueling and that, depending upon how you drive, that number could be as low as 100 miles? How about if she added that a refueling stop would require, not a few minutes at the self-serve, but 3½ hours at a charging station?

Power is defined as energy produced or consumed per unit of time: $P = E/t$. Rearranging that definition gives us the equation $E = P \cdot t$, in which energy equals power times time. A light bulb that is rated at 100 watts, for example, uses energy at that rate: 100 watts. If illuminated for 10 hours, that bulb will consume $100\,\text{W} \times 10\,\text{h}$, or 1,000 watt-hours (1 kilowatt-hour) of energy.* The light bulbs in your home consume electrical energy that is being created continuously by your electric utility (by burning coal or natural gas, or from hydroelectric power, wind power, nuclear energy, etc.) through a distribution grid.

Your home, being a stationary edifice, does not need to store energy on board the same way a wheeled vehicle like a car does. But who knows? Electric cars may someday be able to draw energy from the grid while driving—electric subways and trains do that today. For now, however, an electric car has to store its energy on board, and doing so means batteries.

So let's design an electric car. First on our to-do list: how much energy do we need to store on board? A reasonable efficiency number for an electric car is about 300 watt-hours of battery energy consumed per mile travelled (watt-hours per mile is the electric car analog to miles per gallon of gasoline). If you want a range of 200 miles, you will need batteries that can supply 60 kilowatt-hours of energy (300 watt-hours/mile × 200 miles). Batteries can't be drained completely to zero energy (unlike gasoline tanks), so let's say we end up with a requirement of 70 kilowatt-hours of energy storage.

How much is a battery like that going to weigh? The battery of choice these days in electric vehicles (Tesla Roadster, Chevy Volt, and so on) utilizes lithium-ion technology—similar to the battery technology used in your laptop, cell phone, or iPod. A typical energy density for these types of batteries is an impressive 125 watt-hours per kilogram. That is roughly four times as much energy per kilo as you can store in a traditional lead-acid battery, the starter battery under the hood of most conventional gasoline cars. A 70-kilowatt-hour battery pack, made from 125 watt-hour/kilogram batteries,

*The kilowatt-hour is thus just a convenient unit of energy, equal to exactly 3,600,000 joules, or about 3,412 Btu.

will weigh 70,000 / 125, or 560 kilograms (1,232 pounds). Why is the Tesla Roadster's battery lighter than that? Simple: the Roadster, depending on how you drive it, is more efficient than 300 watt-hours per mile. It's small and ultralight (about 2,700 pounds, and 900 of those pounds are batteries), and it is highly aerodynamic.

In any event, that's what the electric car is up against. In some ways, the gasoline-powered car is just too good. Having won the initial competition against electric cars more than 100 years ago, gasoline cars have now benefited from over a century of intensive development, fueled by trillions of dollars of investment. The result: billions of consumers worldwide who expect—demand—exceptional range and instant refueling, among lots of other things.

Will electric cars finally hit the mainstream, or will they forever remain tomorrow's dream? There is a range of uncertainty in the answer to that question.

16. The Bends

$$p = k_H c$$

Henry's Law

At a given temperature, the solubility of a gas in a liquid is proportional to the partial pressure of that gas above the liquid. Mathematically, the concentration, c, of a gas that is dissolved in a liquid is proportional to the pressure, p, of that particular gas that is in contact with the liquid. The Henry's law constant, k_H, depends on the liquid, the gas, and the temperature.

The Brooklyn Bridge, completed in 1883, was the first to connect Manhattan to Brooklyn. The bridge took over 13 years to build and cost more than $15 million dollars—or about $340 million in 2010 dollars. It was 50% longer than any suspension bridge that had yet been built. And 27 workers died during its construction, many from a mysterious illness dubbed "caisson disease." Today, we call it decompression sickness or simply "the bends."

Among the many technical challenges that had to be overcome, one of the most formidable was the construction of the two iconic towers in the river, one on the Manhattan side, the other near Brooklyn. Soaring 276 feet above the water, these towers were, when they were first built, far taller than just about anything else in the city—or indeed anywhere in America. Anchoring their massive foundations securely in the riverbed required feats of engineering and construction never before attempted on such a grand scale.

To build the towers' foundations, a structure called a caisson was employed. A caisson is essentially an enormous open box turned upside down, so that the open side is on the bottom. The caisson is constructed on shore, floated like a boat to the correct location and then inverted and sunk. It descends to the riverbed, where its edges begin to cut down into the mud and muck and are forced ever downward by the weight of stones placed on top of the caisson by workers. Those stones become the structure of the tower. As the caisson descends through the riverbed, it eventually strikes bedrock.

At that point, concrete is pumped into the inside of the caisson, filling the inverted box, and the tower's foundation is complete.

In theory, that's how it works. In practice, especially for a job as large as that of the Brooklyn Bridge, it is a truly monumental undertaking, fraught with peril at every turn. The Brooklyn Bridge caissons weighed six million pounds each (the weight of 2,000 full-sized sedans). The interior space underneath the caisson down at the bottom of the river, where the construction workers worked, was more than large enough to contain four tennis courts side by side. Each caisson contained 110,000 cubic feet of timber and 230 tons of iron. A schematic diagram of a caisson, in place on the riverbed, is shown in figure 6.

And just what did the workers inside the caisson do? In order for the caisson to make its way downwards towards bedrock, it was necessary to evacuate the mud and rocks that the descending caisson was pushing its way through. To accomplish this, two seven-foot-square vertical shafts, called water shafts, were cut through the roof of each caisson. These hollow steel shafts descended through the open interior of the caisson nearly to the riverbed. Inside the caisson, workers moved rock and mud underneath the bottom of the water shafts, where it was removed from the caisson by huge mechanical shovels lowered from the surface.

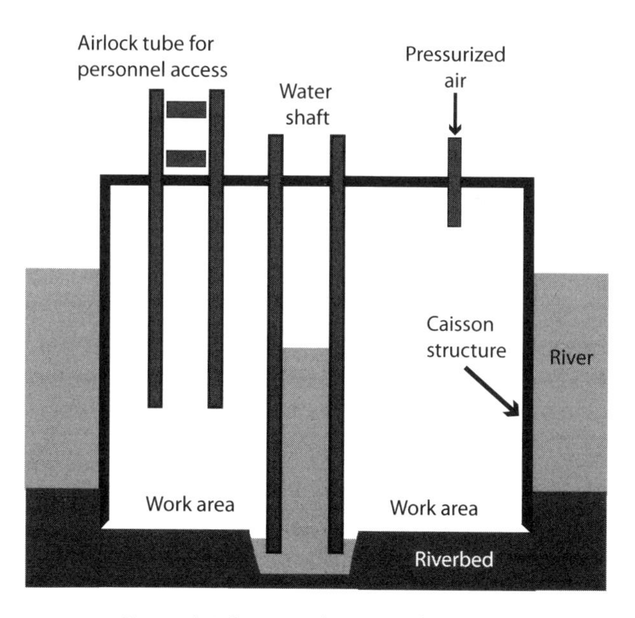

Figure 6. Schematic diagram of a caisson

Making all this work possible meant that the inside of the caisson had to constantly be filled with pressurized air. Otherwise, mud and water from the river would force its way up into the caisson. An air pressure of about 45 pounds per square inch was required—about three times normal atmospheric pressure. To keep this air pressure from simply escaping up the water shafts, the shafts were, as their name suggests, filled with water. It was an ingenious idea. The water in the water shafts was kept at just the right height to balance the elevated air pressure inside the caisson and allow the mechanical shovel to descend through the water in the shaft and remove the debris from inside the caisson. The water shafts thus acted a little like giant barometers. Each increase in air pressure of one pound per square inch inside the caisson required about 27 inches of water in the water shaft to balance it.

Workers entered and exited the caisson through an airlock. This was a steel vessel built into the roof of the caisson with an airtight hatch on the top and another on the bottom. Up to a dozen workers entered the airlock from above, the top hatch was closed, and then a valve was slowly opened, raising the pressure inside the airlock to that of the caisson. At that point, the bottom hatch could be safely opened, allowing the workers to descend into the caisson.

It was there, inside that massive caisson and under three atmospheres of pressure, that their troubles began. And it is also here that we finally get back to our equation, which is known as Henry's law, named for the English chemist William Henry (1774–1836). Henry's law states that the concentration, c, of a gas in a liquid is proportional to the pressure, p, of that particular gas in contact with the liquid. Think of a can of soda pop. When you open the can, the hissing sound you hear is caused by the escaping gases. Keeping the soda under pressure (a high value of p) allows it to retain its carbonation—the fizzy gases dissolved in the beverage (a high value of c). Pour the soda into a glass, and one by one the bubbles of carbonation will make their way to the surface and disappear into the atmosphere, where the concentration of CO_2 is quite low (a low value of p). Within a few hours virtually all of the fizzy bubbles will be gone, and the soda will be flat. Henry's law says that the amount of CO_2 dissolved in the soda—the carbonation level—is proportional to the pressure of CO_2 gas in contact with the soda.

But the liquid we are interested in here is blood, not soda pop. The blood of the caisson workers. And the gas is the nitrogen in the high-pressure air inside the caisson. The high pressure of the air inside the caisson resulted, per Henry's law, in high nitrogen concentrations in the blood of the caisson workers. When those workers left the caisson, it was a little like popping the top on a soda can. The excess nitrogen in their blood came out of solution in the form of gas bubbles inside their blood vessels. This caused excruciating pain, temporary or permanent disability, or death. Individual workers responded very differently, and a lot depended on how long they had been in the caisson.

"Caisson disease" was not new to the Brooklyn Bridge. It had killed more than a dozen workers at the Eads Bridge in St. Louis, which was completed in 1874. And caisson disease was as mysterious as it was deadly. Hypotheses abounded as to what caused the disease and how to prevent it. Caisson workers, not surprisingly, were both terrified of the disease and angry with their management for what were clearly hazardous working conditions. But the workers weren't the only ones who suffered. The bridge's chief engineer, Washington Roebling, made countless trips into the caissons and suffered several bouts of the disease, which eventually left him permanently disabled.

Today, decompression sickness is both well understood and easily prevented. The key is to avoid a rapid change in pressure. Scuba divers, for example, consult detailed tables prescribing just how slowly they must ascend at the end of a dive to avoid creating nitrogen bubbles in their blood. More than 125 years after those famous towers were constructed, the caissons of the Brooklyn Bridge continue to serve both as a strong foundation, under the mud of the East River, and as a hidden monument to those who died in their construction.

17. It's Not the Heat, It's the Humidity

$$°F = 1.8(°C) + 32$$

Relationship between the Fahrenheit and Celsius Temperature Scales

Absolute zero, the lowest possible temperature, occurs at −459.67°F or −273.15°C. The zero points on both the Fahrenheit and Celsius scales thus occur at arbitrary (albeit different) points: 0°C = 32°F and 0°F ≈ −17.8°C. At 40 degrees below zero, the two scales cross: −40°F = −40°C (see figure 7).

To retailers, it is known as psychological pricing. A gallon of gasoline is never priced at $4.00, but rather at $3.99. Psychologically, the latter seems a lot cheaper than the former. And so it is with temperatures. The sweltering summer of 2011 broke plenty of heat records across the United States. Here in Oklahoma, we stayed busy keeping track of how many 100°F days we had. Although we came close, we didn't break the record of 65 days at 100°F or above, which was set in 1936.

But what's so special about 100°F? A temperature of 100°F is hotter than 99°F by exactly as much as 99°F is hotter than 98°F. But 100 has three digits, whereas 99 has only two. And so weather forecasters and others go on endlessly about "another triple digit day, with no end in sight." The point at which two digits becomes three on the Fahrenheit temperature scale is just as arbitrary as any other point on that scale. On the much more widely employed Celsius temperature scale, 100°F is equivalent to about 37.8°C, per our conversion equation above. But in countries that employ the Celsius scale (all countries except the United States, Belize, and a few others), you won't find anyone counting the number of summer days of 37.8°C weather.

Temperature is a slipperier concept than you might imagine, especially when you try to define it at the most basic level. At the everyday level of concern to most of us, temperature is simply a way to quantify hot and cold. One

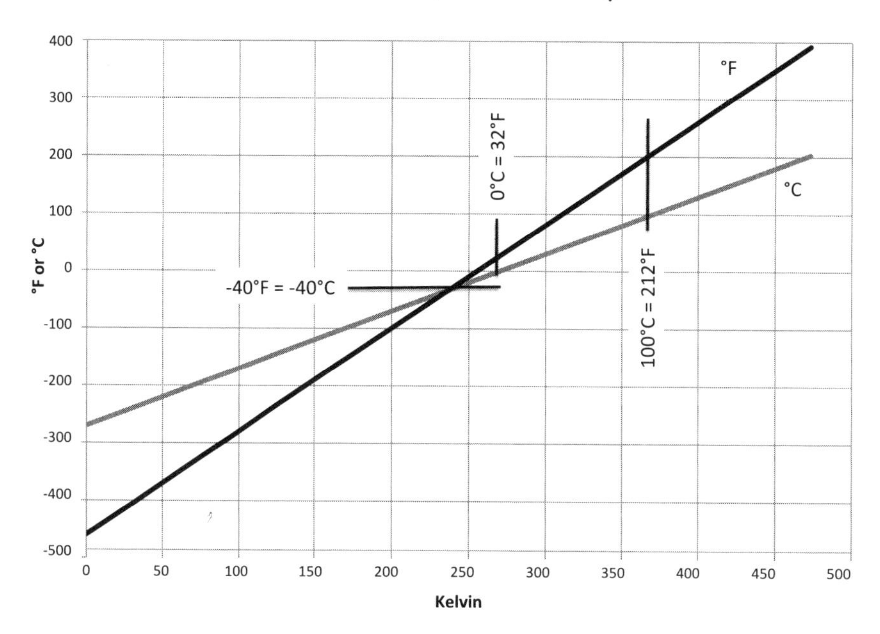

Figure 7. Graphic comparison of the Fahrenheit and Celsius temperature scales on the vertical axis with the absolute temperature, in Kelvins, on the horizontal axis

way that temperature is defined at a more fundamental level relates to the vibration of atoms. All atoms, such as the H_2O molecules in a glass of water, are constantly vibrating. The more they vibrate, the hotter they are. Put the glass of water in a microwave oven for 30 seconds and then insert your finger into the water. The atoms in the water molecules are vibrating faster, as your finger will tell you. Now set the glass in the refrigerator for a half hour or so. Your finger lets you know that the molecules are vibrating more slowly: the water is colder.

Your finger is acting as a thermometer here. It is quite sensitive at this task, allowing you to discern temperature changes of less than 1°F. But it is only a qualitative instrument. It is good at determining relative temperatures—which glass of water is the warmer—but not at quantifying them.

Galileo was one of the first scientists to find a way to link changes in temperature to quantitative phenomena. You can still buy something called a "Galileo thermometer,"* which is a sealed, water-filled glass tube containing a

*There is some doubt as to whether Galileo himself ever constructed one of these gadgets that now bear his name.

set of hollow glass balls filled with colored liquids (the coloring is for aesthetic effect). Each ball has a small calibrated weight suspended from its bottom. A temperature is engraved on each weight. As the temperature of the water in the tube rises, the water becomes less dense, and one by one, the balls inside the tube sink to the bottom. The bottommost ball that is still floating is closest to the temperature of the water in the tube.

The physical principle of the Galileo thermometer is based on the density changes that water undergoes as it heated or cooled. Not too long after Galileo, other scientists employed a related physical phenomenon, that of the thermal expansion of liquids, to create thermometers that would remain the standard until relatively recently. Most liquids expand relatively uniformly as they are heated. Thus, a quantity of liquid trapped in a narrow tube will rise and fall as it is heated and cooled. Alcohol was among the first liquids used for this, and it is still used today—with a colorant added to improve visibility. The metal mercury, now used in thermometers, is one of the few chemical elements that is a liquid at room temperature. The mercury thermometer has been perhaps the most successful in history. It was invented in 1714 by Daniel Gabriel Fahrenheit (1686–1736), a Dutch-Polish physicist.

A few years later, in 1724, Fahrenheit proposed the temperature scale that, with a few minor changes, still bears his name. A variety of explanations have been proposed for the way in which he determined the numerical values for his scale. One version holds that Fahrenheit set 0 and 100 to correspond to the hottest and coldest atmospheric temperatures typically experienced in Western Europe. The truth is somewhat less prosaic. Fahrenheit's scale was built on the work of the Dutch astronomer Ole Rømer, who had himself invented a temperature scale somewhat earlier. For zero degrees on his scale, Fahrenheit chose the temperature of a mixture of ice, water, and ammonium chloride. This is an example of a "frigorific mixture," a concoction whose temperature does not vary with the relative amounts of its ingredients. An ice/water bath is another such frigorific mixture—its temperature does not vary regardless of how little or how much ice is in the mixture. Fahrenheit chose to call the ice/water temperature 32 degrees. He set a third point at 96 degrees, to correspond to the value commonly accepted for the core temperature of the human body. (If these numbers seem inconvenient and haphazard, consider this: the difference between 0 and 32 is exactly one-half the difference between 32 and 96.) Slight modifications made by others eventually resulted in a scale wherein water boils at 212°F and freezes at 32°F, while the core temperature of a typical person lies at about 98.6°F.

Fahrenheit's scale was the standard in many countries as late as the 1960s. It has now been replaced almost everywhere, with the significant exception of the United States, by the Celsius scale, named for the Swedish astronomer Anders Celsius (1701–44). On the Celsius scale, water freezes at 0°C and, at atmospheric pressure, boils at 100°C. The more practical nature of the Celsius scale scarcely needs to be pointed out. Its success speaks for itself. Why it has not been universally adopted in the United States remains a mystery. Perhaps it is, at least in part, so that we can continue to count the number of days over 100°F during a long, hot summer.

18. The World's Most Beautiful Equation

$$e^{i\pi} + 1 = 0$$

Euler's Identity

Euler's identity states that the constant e raised to the power $i\pi$, plus 1, equals zero. The value of i is the square root of -1. Euler's identity is a special case of Euler's formula:

$$e^{ix} = \cos(x) + i\sin(x).$$

When $x = \pi$, Euler's formula becomes Euler's identity, since $\cos(\pi) = -1$ and $\sin(\pi) = 0$.

In 2004, the journal *Physics World* asked readers to nominate the "twenty most beautiful equations in science." Our equation here, often called Euler's identity, tied for first place for most nominations. A similar poll taken by the *Mathematical Intelligencer* in 1990 identified Euler's identity as "the most beautiful theorem in mathematics."

But beauty, as they say, is in the eye of the beholder. Is the *Mona Lisa* a more beautiful painting than Van Gogh's *Starry Night*? Is the Grand Canyon more beautiful than Niagara Falls? We can never say for sure; beauty is just too difficult to quantify, thank goodness. So what is it about Euler's identity that inspires so many folks to extoll its beauty?

First off, let's note that our equation here is a little different from those in most of our other stories. Generally, the term "equation" implies something that can be solved. For example, we can solve the equation $x + 2 = 4$ and get the result that $x = 2$. But Euler's identity isn't like that. There's nothing to solve for. It's just a statement of fact, like saying that $2 + 2 = 4$. And so just as $2 + 2 = 4$ is a fact, $e^{i\pi} + 1 = 0$ is also a fact. What makes the latter beautiful and the former banal?

Mathematician and author David Wells, who proposed the poll in the *Mathematical Intelligencer* noted above, provides a helpful list for anyone trying to decide if any particular mathematical expression is beautiful. To be beautiful, according to Wells, an expression must be simple, brief, important, and surprising. Our first example, $2+2=4$, is simple and brief but not particularly important or surprising.

Is Euler's identity simple? That depends on whom you ask. Is it brief? Absolutely. Important? Undoubtedly (as discussed below). Surprising? Why yes, as a matter of fact it *is* surprising!

Euler's identity is surprising for a variety of mathematical reasons. It is surprising to think that you can raise a number to some power such the result ends up being less than zero. This, to say the least, is unusual. Others have pointed out that Euler's identity is surprising (not to mention quite elegant) because it contains exactly one addition, one multiplication, and one exponentiation. It also contains, once each, perhaps the two most famous and important mathematical constants, e (the base of the natural logarithm) and π (the ratio of a circle's circumference to its diameter). It also contains, once each, the additive identity (0) and the multiplicative identity (1). Finally, it contains the mysterious imaginary number i, which equals the square root of -1.

Does all of this add up to an unrivalled mathematical beauty? We'll leave that one up to you, dear reader.

Euler's identity may be renowned for its beauty, but the more general Euler's formula,

$$e^{ix} = \cos(x) + i\sin(x)$$

is all business. The formula is of substantial import not only in the realm of pure mathematics but also in science and engineering, particularly electrical engineering. The analysis of alternating current circuits, for example, relies heavily on Euler's e^{ix} formula.

There is some evidence that Euler wasn't the first to discover the identity that bears his name, and it is probably true that he never wrote out the identity exactly as we have represented it in this story. No matter. Eulerian fingerprints are all over the math contained therein—a fact few would dispute. Look where you will throughout the length and breadth of any branch in the field of mathematics. Chances are you will stumble across Euler's name sooner rather than later.

Swiss mathematician Leonhard Euler (1707–83) is on nearly everyone's list of the five greatest mathematicians of all time.* "The greatest" is no easier

*The other names on such lists nearly always include Newton, Gauss, and Archimedes.

to quantify than "the most beautiful," but please bear with us anyway. Euler was both brilliant and prolific—an unbeatable combination in nearly any field. Several other stories in this book owe a great debt to Mr. Euler. With a little more effort, we might have written a whole book of stories about nothing but Euler's equations.

Euler's contributions as a mathematician and mathematical physicist touch on areas too numerous to mention. Just the list of things mathematical and physical named after Euler runs to several pages. These include Euler equations, Euler formulas, Euler identities, Euler theorems, Euler functions, Euler numbers, Euler geometric forms, Euler laws, and Euler conjectures, among other things.

Euler suffered from multiple vision problems and was profoundly blind for the last 17 years of his life. If anything, he was more prolific as a blind man than he had been when he could see. In 1775, nine years after becoming blind, he produced, on average, one mathematical paper per week. Fifty-two such papers many a career would make; Euler polished them off in a single year. Euler possessed phenomenal gifts for mental calculation—crunching numbers in his head—and his memory was equally amazing. This allowed an old blind man to create new math in his head in real time and to recite it orally for colleagues, who scribbled furiously to keep up with him. When he died, one eulogy noted that, "he ceased to calculate, and to live." For Euler, life and calculation were nearly one and the same.

19. Breaking the Law

$$\Delta E_{\text{int}} = Q - W$$

A Representation of the First Law of Thermodynamics

According to the first law of thermodynamics, which is sometimes referred to as the law of conservation of energy, the energy of a closed system can be transformed from one form to another, but the total amount of energy cannot be increased or decreased. Mathematically, the change in internal energy of a closed system, ΔE_{int}, is equal to the energy transferred by heat to the system, Q, minus the work done by the system, W.

In 2006, the Irish company Steorn placed an advertisement in the *Economist* claiming to have created a device that produced energy that was "free, clean, and constant." Such a claim, if true, would violate the first law of thermodynamics. A team of skeptical scientists who formed, at Steorn's request, a jury to evaluate the company's claim quickly concluded that Steorn had done no such thing. Several public demonstrations by Steorn since then have done little to advance the company's cause.

Why is it such a big deal to break the first law of thermodynamics, or even to contemplate doing so? For one thing, it has never been done. The first law of thermodynamics, which was formulated by the German physicist Rudolf Clausius in 1850, is essentially a statement of the law of conservation of energy. If a device exists that violates the first law, we would have likely have on our hands a solution to the world's energy needs. The first law can be represented, mathematically, in a variety of ways. In our equation above, the change in the internal energy of a system, ΔE_{int}, is equal to the energy transferred by heat to the system, Q, minus the work done by the system, W. Simply put, energy can be changed from one form to another, but it cannot be created, nor can it be destroyed.

Energy can exist in various forms. In gasoline, for example, energy is positively crammed into the chemicals bonds of this potent liquid fuel. When gasoline is burned, however, the energy is liberated from those chemical bonds and no longer exists in such a concentrated form. What was once chemical energy has now become heat. It exists in precisely the same quantity that was once contained in the gasoline's chemical bonds but is now present in the surrounding air.

A machine that creates energy would violate the first law of thermodynamics and would also be one kind of perpetual motion machine. Before we go too far, let us note that no one has ever successfully created a perpetual motion machine of *any* sort. That hasn't stopped lots of folks from trying, however, and over the years a system for categorizing perpetual motion machines into three kinds has developed. The first kind of perpetual motion machine is epitomized by the Steorn example: a machine that, once set in motion, continues to do useful work without any input of energy, or one that produces more energy than it consumes. Such a machine, as noted, would violate the first law of thermodynamics.

The second kind of perpetual motion machine is one that would convert heat completely into useful work, with no waste, in violation of the second law of thermodynamics. Have you ever burned your finger on the hot exhaust pipe of a lawnmower, motorcycle, or car? If the engine in question had been a perpetual motion machine of the second kind, no such "waste heat" whatsoever would be rejected by the engine, as it converted the chemical energy in its gasoline fuel entirely into useful work. The third kind of perpetual motion machine is the kind most often associated with this term: one that, due to the elimination of friction and other energy-dissipating mechanisms, remains in motion forever.

Attempts to create perpetual motion machines—and they have been too numerous to count—often try to counteract the first law by taking advantage of leverage, as in the weighted wheel example shown in figure 8, or magnetism, as in the case of Steorn's device. The weighted wheel is based on the concept that differences in leverage will allow the wheel to continue spinning once it is started. The weights on the right are being pulled downward by gravity. With the extension of the lever arms, a falling weight will produce a torque greater than that of a weight rising on the opposite side. There are, however, *more weights* on the rising side, and even though each rising weight produces less torque than each falling weight, the net result is that there is just as much torque resisting the motion of the wheel as there is creating that motion.

Since there will always be some friction in the bearings of the wheel and in the air, energy is lost, albeit slowly, and the wheel eventually comes to a

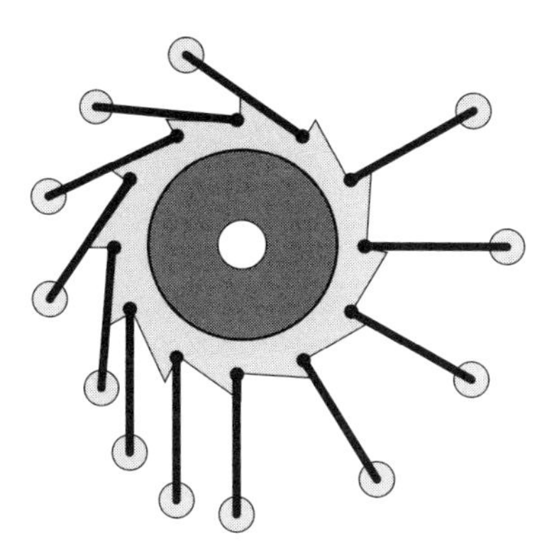

Figure 8. Schematic of a weighted, or overbalanced, wheel

stop. Perpetual motion devices utilize all sorts of scientific principles and come in all different shapes and sizes, but they have one depressing feature in common: none of them has ever worked.

Perpetual motion machines occupy a special place in the world of inventions. The French Academy of Science has refused, since 1775, to communicate with anyone claiming to have created a perpetual motion device. In the United States, it has been many years since the U.S. Patent and Trademark Office required all inventors to submit working models along with patent claims, although you must still prove to the satisfaction of the patent examiners that your device performs as advertised. Should you desire to patent a perpetual motion machine, however, your attention is directed to the Patent Office's *Manual of Patent Examining Procedure*, section 608.03: "With the exception of cases involving perpetual motion, a model is not ordinarily required by the Office to demonstrate the operability of a device."

The Patent Office issued patents on perpetual motion machines for over 150 years. At some point it decided that it had had enough, and the above policy went into place. Having been burned by one too many charlatans, the Patent Office is no longer willing to pass judgment on perpetual motion devices passed solely on paper claims. Got a great idea for a perpetual motion machine? Build it, and then the patent folks will take a look at it.

Physicists, a serious lot, tend to shy away from terms like "impossible." Perpetual motion, they will tell you, is neither possible nor impossible. It simply violates the laws of physics as we now understand them. Will those laws ever be broken, and thus rewritten? Perhaps. A fellow named Einstein changed our understanding of physical laws in fundamental ways, in effect rewriting some of the laws of physics. But not the first law of thermodynamics.

20. The Mars Curse

$$1 \text{ lb}_f \approx 4.448 \text{ N}$$

Units Conversion from Pounds-Force to Newtons

This equation represents the approximate conversion between U.S. customary units for force (the pound-force or lb_f) and the SI (metric system) unit the newton (N). One newton is the amount of force necessary to accelerate a mass of one kilogram at the rate of one meter per second squared. In equation form: $1 \text{ N} = 1 \text{ kg}(1 \text{ m/s}^2)$. In the U.S. customary system, the pound-force appears in the following fundamental relationship: $1 \text{ slug} = 1 \text{ lb}_f(\text{s}^2/\text{ft})$. In this system, the slug is the fundamental unit of mass, as derived from the pound-force, the foot, and the second.

Imagine that you are driving your American car, with its American speedometer, in a foreign country. On a winding mountain road, a road sign indicates a curve to the left along with a large "60." You adjust your speed to 60 miles per hour as you enter the curve. Suddenly, you realize that the curve is much too sharp for that speed, and you slam on the brakes, praying that you don't go skidding off the road to the right and into the valley below. The road sign signified 60 kilometers per hour and was warning you not to exceed the equivalent speed of about 37 miles per hour. The road sign and your speedometer were both communicating with you, the driver, but unfortunately they weren't speaking the same language.

If you were lucky enough to survive such a harrowing journey, you would surely notice that the steering and braking corrections you had to make were much more severe than those that would have been required had your speedometer been calibrated, along with the road signs, in the same units of measure. Although you never left the planet Earth, one might say that you were suffering from the Mars Curse.

Although Venus is the closest planet to the Earth, there can be no doubt that we Earthlings are, in general, much more fascinated by the red planet known as Mars. Ever since technology has allowed humans to send spacecraft out into the unknown, the siren call of Mars has been hard to resist. And since the Soviets first attempted to reach the planet in 1960, there have been at least 42 unmanned vehicles designed to send data on Mars back to Earth either by flying by the planet, orbiting it, or landing on its surface. Unfortunately, half, or more, of those missions ended in abject failure and none more infamously than the one featuring the *Mars Climate Orbiter* in 1999. That mission gave rise to what has been called the Mars Curse.

The *Mars Climate Orbiter*, or *MCO*, was a 750-pound spacecraft designed to orbit Mars and gather data about the planet's atmosphere and climate in preparation for future missions. It was billed by NASA as the first "interplanetary weather satellite." The *MCO* was launched in December 1998. In September 1999, the mission failed when the craft destroyed itself upon entering the Martian atmosphere at a wildly improper altitude.

If the *Mars Climate Orbiter*'s computer brain could have spoken, it might have remarked, before the mission was lost, "Houston, we have a problem . . . with units conversion." For it was as simple as that. Software on board the *MCO* controlled how and when the spacecraft's thruster rockets would fire as the vessel made its nine-month journey from the Earth to Mars, and then, crucially, as the *MCO* transitioned from space to the Martian atmosphere. The manner in which a space vehicle enters the atmosphere of a planet is vitally important. Moving at high speed (on the order of 10,000 miles per hour) from the vacuum of outer space to the density of a gaseous atmosphere requires that a spacecraft follow a carefully chosen path in order to minimize the stresses induced on the craft. To follow such a path, a spacecraft must fire its thrusters at just the right time and with just the right intensity. Otherwise the mechanical and thermal stresses induced by the atmosphere on the spacecraft will be too severe.

An unmanned spacecraft is controlled by onboard software that follows commands sent by the mission's commanders on Earth. To keep it on the correct path, the *MCO* had nine different thrusters (small rockets). Four of the thrusters could produce about five pounds of force (about 22.2 N), while four others were much smaller still, with a maximum thrust of only about two-tenths of a pound (about 0.9 N). These eight thrusters took care of the direction the spacecraft was travelling through space on the long trip to Mars, and also the *MCO*'s attitude, which refers to its orientation in space. The ninth thruster was much larger, capable of providing about 144 pounds of thrust (about 640.5 N). This larger thruster's only job was to help ensure that the

MCO was safely "inserted" into the Martian atmosphere, so that it would achieve a stable orbit at the correct altitude.

Throughout the *MCO*'s voyage to Mars, mission commanders on Earth noticed a perplexing problem: they had to make more corrections than usual to the spacecraft's trajectory and attitude. This was due in part to the design of the *MCO*, but it was also due to a communications blunder that would have made a ninth-grade science student blush.

Specifically, and not unlike our mountain driving anecdote, the commands sent from Earth to the *MCO* utilized U.S. customary units, while the software on the *MCO* was written for SI units, or what is commonly called the metric system. The U.S. customary unit for force is the pound-force, or lb_f.* The SI unit for force is called a newton. One pound-force is equal to about 4.45 newtons—our equation here. The commands being sent to the thrusters on the *MCO* were thus off by a factor of 4.45. The result was that as the *MCO* approached Mars for its initial insertion into the atmosphere, it was far too low—far too close to the planet's surface. What should have been an altitude of about 140 miles was instead more like 35 miles. Entering the Martian atmosphere at that altitude was something the spacecraft could not survive.

NASA's report on the mishap includes the following recommendation for future missions: "Verify the consistent use of units throughout the spacecraft design and operations." Obvious? Yes, but someone did have to say it.

In the years since the *Mars Climate Orbiter* mission, the Mars Curse seems to have loosened its grip on those of us here on Earth seeking to unlock the secrets of the Red Planet. The *Spirit* and *Opportunity* rovers, for example, roamed the Martian surface for far longer than the most optimistic projections, and sent back a treasure trove of digital information.

Will we someday send humans to Mars? Perhaps we will, although the challenges of such a mission are extreme, extending, as they do, far above and beyond the rather mundane requirement of making sure that the units of measurement are consistent.

*In everyday language, when Americans refer to a "pound," they are referring to the pound-force.

21. Eureka!

$$\rho = m/V$$

Equation for the Density of an Object

The density, ρ, of a quantity of matter is equal to its mass, m, divided by its volume, V. The SI units for density are kilograms per meter cubed (kg/m^3). The density of pure water is often taken to be 1,000 kg/m^3 (1 kg/liter), although this is only an approximation. At its densest (at 4°C) water has a density of about 999.97 kg/m^3. The specific gravity of a substance is its density divided by the density of water. A specific gravity less than 1 means a substance is less dense than water.

Did Archimedes really run naked through the streets of Syracuse shouting "Eureka!" when he discovered the principle of buoyancy? We'll never know with certainty, but it sure makes for a great story.

Tasked with determining whether King Hiero II was being cheated by his goldsmith, Archimedes was given a golden crown to evaluate. Archimedes knew that one way to determine whether the crown was pure gold was to compare its density to the known density of gold. If the crown could be shown to be less dense than pure gold, then it had obviously been alloyed with cheaper, lighter elements such as copper. Archimedes knew, as well, that the density of an object followed our equation here, $\rho = m/V$, where ρ is the density of an object, m is its mass, and V its volume. Finding the mass of the crown was simple enough back then, but the intricate laurel wreath shape of the crown made finding its volume problematic.

As the story goes, Archimedes was contemplating all this one fine day in his bathtub, when he hit upon a solution. He realized that when he immersed himself in the tub, the water level went up by an amount equal to the volume of his own body. If he could measure the increase in the water level, he would know the volume of his body—or of any other body he immersed in water, including the king's new crown. And thus, the original Eureka moment, followed shortly thereafter by a famous mathematician streaking through his neighborhood, shouting, "I have found it!" (from the ancient Greek *Eureka*).

The technique just described works great in theory. But it has often been pointed out that in practice, it would have been quite difficult to use the water displacement method to get an accurate enough measure of the volume of the crown to make a useful estimate of its density.

The "Eureka" story comes to us not from Archimedes himself but from the Roman writer, architect, and engineer Vitruvius, as recounted in his monumental work *De Architectura*. Archimedes makes no mention of the story of King Hiero's crown in his own writings. He did, however, write about the principle of buoyancy, which is today sometimes referred to as Archimedes' principle, in his treatise *On Floating Bodies*. Archimedes' buoyancy principle states that when an object is wholly or partially immersed in a fluid, the object is pushed upwards by a force equal to the weight of the fluid that the object displaced.

This principle could have been used to solve the crown mystery as follows. Using a two-pan balance (à la the Scales of Justice), put the crown on one pan and balance it with an equal mass of pure gold on the other. Now, lower the balance into a container of water, such that the crown and the gold are completely immersed. If the crown is pure gold, it will remain in balance with the pan containing the pure gold. If the crown is made of a lighter alloy, however, it will have a greater volume than the pure gold on the other side of the balance, and thus the buoyant force pushing the objects upwards will be greater on the crown than on the pure gold. Under water, the crown will be out of balance with the pure gold, and the crown will float up to a higher level. This water balance technique is very sensitive and does not require the measurement of the volume of water displaced, as is necessary in the Eureka story. For this reason, no less an authority than Galileo (who came along centuries later) is said to have concluded that Archimedes probably used the water balance technique to evaluate King Hiero's crown.

Fluid balances (often using liquids other than water) are still used today to evaluate very small density differences in materials such as plastics. And the principle of buoyancy is at work every time a ship goes to sea. Large ocean-going vessels are nearly always made of steel, which is nearly eight times as dense as the water the ships must float on. Such ships float on water because the steel hull traps a large volume of air inside it. The density of the ship—the mass of the steel plus air divided by the total volume—is less than the density of the water.

Archimedes is on most lists of the greatest mathematicians of all time and is considered the greatest mathematician of antiquity. His contributions were scarcely limited to the realm of pure mathematics, however. As a physicist, he laid the foundations for the modern fields of statics and hydraulics. For the former, there is the principle of the lever, and for the latter, an efficient pump we now call the Archimedes screw. He is also often credited with a weapon of war sometimes called Archimedes heat ray. This device uses a large number of mirrors to focus sunlight on a single object, such as an enemy's wooden ship approaching the shore. If enough solar energy is focused on it, wood can be made to burst into flames. This technique has been subjected to countless modern recreations, with mixed results, including several famous ones on the popular television program *MythBusters* (wherein the myth was busted).

Archimedes died in 212 BCE, at the hands of a Roman soldier. Syracuse, during the Second Punic War, had been under siege for two years. When the city finally fell, Roman soldiers were under strict orders to take Archimedes into custody unharmed. Capturing Archimedes back then would have been roughly equivalent to capturing Albert Einstein during World War II. Plutarch offers several versions of what happened when a Roman soldier burst into Archimedes' quarters. In one such account, Archimedes refuses to accompany the soldier, saying he has work to do, whereupon the soldier kills him. Archimedes' last words, referring to the drawings on which he was working at the time, are said to have been, "Do not disturb my circles."

22. A Penny Saved...

$$FV = PV\left(1+\frac{i}{n}\right)^{Yn}$$

An Equation for Calculating Compound Interest

The future value, FV, of a sum of money is computed from its present value, PV. The interest rate is i, the number of years is Y, and n is the number of times the interest is compounded per year. In the case of continuously compounded interest, n approaches infinity, and the equation becomes $FV = PVe^{Yi}$.

Back in the good old days, folks used to keep their money down at the local bank, in a savings account. You gave the bank your money, which the bank could then lend to other people, and in exchange it paid you a little bit of extra money for your trouble. That little bit of extra cash is called interest.

Let's say that you deposit $100 in your savings account and that the bank is willing to pay you 5% interest, compounded annually. How much money will you have at the end of one year? In our equation above, PV is the present value of your money, $100 in this case. The equation calculates the future value, FV, given an interest rate of i; the number of times, n, that the interest is compounded annually; and the number of years, Y. In our simple example, $i = 5\%$, $n = 1$, and $Y = 1$. Thus, the future value of your $100 is $105, after one year's interest at 5%.

Interest also works in the other direction. If you wanted to borrow that same $100 from the bank, it would lend the money to you—with interest. You could use the equation above to figure out how much extra money, above and beyond what you borrowed, you would owe the bank for lending you the money for a certain amount of time at a given interest rate.

Questions and calculations like these are so integral to our modern economic system that it's hard to imagine that things weren't always that way, or that there are cultures even today where things don't work quite the same way. We pay interest on home loans, car loans, and credit card debt. We receive interest on money deposited in savings accounts or bonds. But did you know, for example, that the Bible forbids charging interest on a loan?

It's right there in the Old Testament, Deuteronomy 23:19, in the New American Standard Bible: "You shall not charge interest to your countrymen: interest on money, food, or anything that may be loaned at interest." For a very long time the Catholic Church condemned the charging of interest. Saint Thomas Aquinas (1225–74) held that charging interest on a loan was wrong because, in doing so, you are charging both for "the thing" and for "the use of the thing"—and thus double charging. In those days the term "usury" was used whenever any sort of interest was charged on anything. Today, usury refers only to exorbitant interest rates.

Eventually, Western thought came around to the idea that charging interest was a reasonable way of doing business. In the late 1400s and early 1500s, the School of Salamanca in Spain was an influential center of ideas regarding diverse topics such as theology, justice, and economics. The basic ideas, if not the specific equations, of the time value of money were elucidated there. When someone lends you money, he is being deprived of the use of that money throughout the period of the loan. To be compensated for that deprivation is only natural, the Salamancans argued. Thus, it's okay to charge interest. In 1545, under King Henry VIII, the British passed a law legalizing the charging of interest.

Even today, however, the concept of interest can be problematic in the Islamic world. Islamic banks are required to comply with Islamic law, which generally prohibits the charging of interest. A Western bank will lend you money to buy a house, for example. This is called a mortgage loan, and the buyer must repay the bank with interest over the term of the loan, which includes strict penalties for late payments. To avoid charging interest, an Islamic bank might buy a home and then sell it to you at a profit. You would then repay the bank in installments, although the bank is forbidden to charge penalties for late payments. To protect itself against default, the bank requires collateral for all such transactions. It might seem as if both banks, Western and Islamic, end up in the same place here, since both make a profit in helping you to buy a house, but there are differences. These days, big international corporations wishing to do business in Muslim countries often invest a lot of time and money training their personnel and establishing business practices that will be acceptable in the Islamic world.

Religious questions aside, the mathematics of interest rates, compounding periods, and the like might seem to rest firmly in the realm of practical, applied mathematics. Such calculations, however, do indeed contain principles important in the world of "pure" mathematics.

The mathematical constant e is not particularly well known to the general public, but to scientists and mathematicians it is at least as important as its renowned cousin pi (π). The constant e (the notation is due to the prolific and incomparable Leonhard Euler) can be defined various ways. It is, for example, the number whose natural logarithm is exactly 1. One of the first persons to stumble upon the constant e was the Swiss mathematician Jacob Bernoulli, who, in 1683, was investigating, not logarithms, but compound interest. In particular, he was studying interest that is compounded continuously.

Before the advent of the digital computer, interest was calculated only a few times a year. On a typical savings account, for example, interest was usually compounded quarterly—four times per year. Modern computers make it easy to compound interest continually. The equation for continuously compounded interest is a little different from our equation above. The future value, FV, of a sum of money currently valued at PV is calculated as shown below for an interest rate of i for Y years:

$$FV = \lim_{n \to \infty} PV \left(1 + \frac{i}{n}\right)^{Yn}.$$

It was the mathematics of this limit that Bernoulli was investigating:

$$\lim_{n \to \infty} \left(1 + \frac{1}{n}\right)^{n} = e \approx 2.718.$$

Bernoulli was only able to show that the limit above lies between 2 and 3. It fell to others to calculate e with greater precision (e, like π, is an irrational number).

In any event, given that the limit above is equal to e, the equation for continuously compounded interest becomes

$$FV = PV\, e^{Yi}.$$

For our example above, the difference between continuously compounded interest and interest compounded annually is tiny: $100 invested at 5% interest becomes $105.00 when compounded annually and only $105.13 when compounded continuously. Don't be fooled, however. The differences over longer periods of time, and for greater interest rates, can be much more significant.

23. If I Only Had a Brain

$$a^2 + b^2 = c^2$$

The Pythagorean Theorem

The Pythagorean theorem relates the lengths of the sides of a right triangle to its hypotenuse. In a right triangle, one of the three angles is equal to 90 degrees. The two shorter sides of the triangle have lengths a and b. Opposite the 90-degree angle, the longest side, the hypotenuse, has length c. The sum of the squares of the lengths of the two short sides, $a^2 + b^2$, is equal to the square of the length of the hypotenuse, c^2.

How best to go about proving your intelligence to your friends? Why not take your cue from the Scarecrow in the movie *The Wizard of Oz*. At the end of the film, the Scarecrow finally gets his brain. To show off his newly acquired smarts, he instantly rattles off the Pythagorean theorem, and then exclaims, "Oh joy, rapture, I've got a brain!" The joke is that the Scarecrow gets it wrong; his version of the Pythagorean theorem is mangled almost beyond recognition. Let's see if we can set our friend straight.

The Pythagorean theorem is one of the best-known and most useful relationships in all of geometry. It applies to all right triangles—triangles that include an angle of 90 degrees. For any right triangle, the length of its longest side, the hypotenuse, when squared, is equal to the sum of the squares of the lengths of the other two sides. In our equation, c is the length of the hypotenuse, and a and b are the lengths of the other two sides.

Pythagoras was probably born on the Greek island of Samos sometime around 570 BCE. He became one of the best known of what we now call the pre-Socratic philosophers of ancient Greece. His interests ranged far beyond mathematics to encompass metaphysics, ethics, politics, and music, but relatively little is known with certainty about his life of about 75 years. Most of what we know (or think we know) about Pythagoras comes from writings that postdate him by several hundred years. He is, and will probably always remain, an enigma.

"Numbers rule the universe." This was the motto of the Pythagoreans, a secret society of scholars led by its namesake. It was their secrecy, combined with their penchant for the oral tradition of passing down what they had learned, that leave us today with such a paucity of knowledge about them. An early discovery attributed to Pythagoras and his society was in the field of acoustics, wherein the musical pitch of a vibrating string was found to be inversely proportional to its length. This discovery marked one of the first—if not *the* first—examples of the quantification of a natural phenomenon, or what we now call mathematical physics. That it came in the area of music is not surprising. Music was one corner of the *quadrivium*, the four core elements of the ancient Greek curriculum, along with arithmetic, geometry, and astronomy. Music was thus on a par with mathematics, an idea which seems a little strange today, although it shouldn't. The Pythagoreans are given credit for lots of other musical discoveries, although once again it's difficult to know with certainty just what they accomplished.

The history of the theorem that bears the Pythagorean name is likewise shrouded in mystery. The theorem was known to the Babylonians, and perhaps to the Chinese, up to a thousand years before Pythagoras. It is possible that Pythagoras and his followers were the first to mathematically prove the theorem. Or not.

Speaking of proof, the Pythagorean theorem is relatively easy to prove mathematically, and it can be proven many different ways, some simple, some dauntingly complex. In fact, few if any mathematical theorems have been proven in so many different ways. More than four hundred different original proofs of the theorem exist, according to Eli Maor, in his book *The Pythagorean Theorem: A 4,000-Year History*. No less a luminary than Albert Einstein wrote a unique proof—as a 12-year-old boy.

Another original proof was set down by James Garfield, later to become president of the United States. Garfield's proof was created in 1876, while its author was a member of the U.S. House of Representatives. Garfield wrote that the proof came to him during, as he put it, "some mathematical amusements and discussions with other M.C. [members of Congress]." Garfield later published his proof in the *New England Journal of Education*. Times have changed, and today, one suspects that moments of "mathematical amusements and discussions" among members of Congress are few and far between. But let's leave politics aside. Mathematics doesn't get much more pure than the Pythagorean theorem.

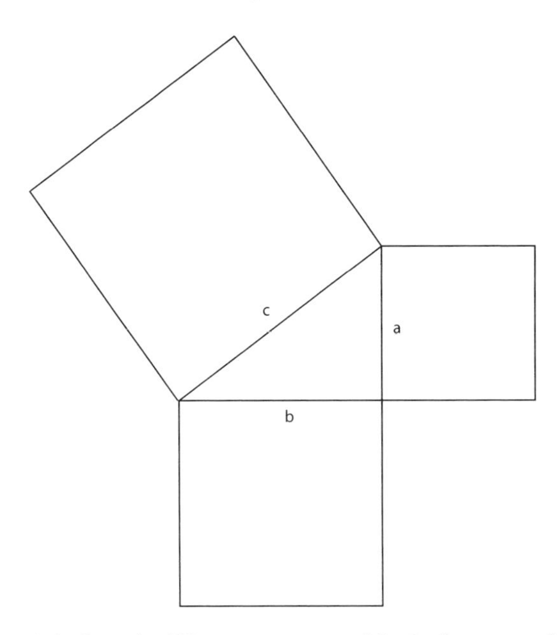

Figure 9. The "windmill" representation of the Pythagorean theorem

Today the Pythagorean theorem is thought of as we described it above: a geometric relationship for right triangles that can be expressed with a nice, neat algebraic statement: $a^2 + b^2 = c^2$. In Pythagorean times and before, however, it was more about areas than it was about triangles or algebra. Figure 9 shows why. This figure, famously associated with the Pythagorean theorem, has been variously referred to as the windmill, the peacock's tail, or the bride's chair. And it is just as correct to say that the area of the larger square (c) is equal to the sum of the areas of the two smaller squares (a and b) as it is to say that $a^2 + b^2 = c^2$. These two ways of looking at the Pythagorean theorem are mathematically equivalent.

Some have suggested that the windmill representation of the Pythagorean theorem be used as an enormous permanent signal (crafted from rows of pine trees and huge wheat fields in Siberia and large enough to be seen from Mars) to let all the extraterrestrials out there watching us know that there really *is* intelligent life down here on Earth. Often attributed (probably erroneously) to the great German mathematician Karl Friedrich Gauss, the plan has never progressed past the idea stage.

The Pythagorean theorem holds for all right triangles. Almost from the beginning, however, mathematicians were extremely keen on studying those cases in which a, b, and c are positive integers (whole numbers greater than zero). For example, a right triangle with sides of 3, 4, and 5 fits the Pythagorean theorem, since $3^2 + 4^2 = 5^2$. There are, in fact, an infinite number of such "Pythagorean triples" in which a, b, and c are positive integers. Put another way, it is not unusual for the square of a whole number (5 in our example above) to be able to be expressed as the sum of the squares of two other numbers (3 and 4 in this case).

But what would happen if, instead of squaring all three numbers, we were raising them to a higher power, such as 3 or 4? It turns out that *there isn't a single example*—for any possible combination of positive whole-number values of a, b, and c—for which this holds true. There are not, for example, any such values of a, b, and c for which $a^3 + b^3 = c^3$. This is a mathematical fact; it is as true as the Pythagorean theorem. But how on earth did anyone ever figure that out? That, as they say, is another story, and we include it here as our very next one.

24. Because It Was There

$$a^n + b^n = c^n$$

The Fermat-Wiles Theorem, or Fermat's Last Theorem

The exponent n is a positive integer, as are a, b, and c. The theorem states that there are no possible solutions to the equation for n greater than 2. The theorem was famously conjectured by Pierre de Fermat in 1637, but not proven until 1994, by Sir Andrew Wiles.

Most of the great mathematical puzzles are pretty much impenetrable to the average person. Not only do we not understand the answers, but we mere mortals have a heck of a time even understanding the questions that the great mathematicians ask and attempt to answer. Not so with our subject here. Fermat's last theorem is different in that the question at hand is quite easy to understand. The proof—for there is now mathematical proof that Fermat's last theorem is true—is dauntingly complex, but the theorem itself is accessible to anyone with a little knowledge of algebra.

We'll get back to Mr. Fermat in just a moment. But first, to see just how simple the above equation is, let n equal 1. Then the equation just says that $a + b = c$. There are obviously an infinite number of cases where this is true, so long as a, b, and c are whole numbers greater than zero (positive integers). For example, $1 + 2 = 3$. Now let n equal 2, and the equation becomes $a^2 + b^2 = c^2$. This is just the Pythagorean theorem, discussed in our previous story. If a, b, and c are positive integers, there are lots of cases where this equation is true. For example, $3^2 + 4^2 = 5^2$, or $5^2 + 12^2 = 13^2$.

So, there are plenty of examples where our equation holds true if n equals 1 or 2. What about if n equals 3? Or 4? Or some larger value? Fermat's last theorem states that *this equation has no solutions*—no positive integer values of a, b, and c for which the equation is true—if n is larger than 2. But that seems crazy, on at least two fronts. First, there are an infinite number of possibilities for a, b, and c, so how could it possibly be that *there isn't a single case* when our equation is true if n is larger than 2? Well, that's Fermat's last theorem. And second, how could anyone possibly be able to show that this is true?

"I've found a marvelous proof of this," Fermat wrote, by hand, in the margins of the Latin translation of a Greek mathematics text that discusses this particular equation. Then, in what is undoubtedly the most famous of all teasers in the history of mathematics, he added, "The margin of the page is too small to include it." Too small, indeed.

For over three hundred years, mathematicians searched in vain for that "marvelous proof." What we now know popularly as Fermat's last theorem is more correctly called the Fermat-Wiles Theorem. It wasn't proven until 1994, by the British mathematician Andrew Wiles. To say that the margin of a single page would be too small to include the proof would be to understate things ever so slightly. Wiles's proof, as published in 1995, runs to more than a hundred pages. Figure 10 shows a stamp from the Czech Republic commemorating the event in 2000.

Pierre de Fermat was born in 1601 and died in 1665. He was a lawyer by trade. As a mathematician, he was essentially a hobbyist, although a brilliant one. Fermat is known in his native France as *le Prince des Amateurs*, which does not quite translate as "the Prince of Amateurs." In French, *amateur* is closer in meaning to "enthusiast" or "connoisseur" in English. Fermat, then, might better be called "the Prince of Math Connoisseurs." He communicated with most of the great mathematicians of his day, but published only a few papers. It is mostly through his written correspondence with his contemporaries that we know of his genius.

Figure 10. A stamp from the Czech Republic in 2000 commemorating Fermat's last theorem and Andrew Wiles's proof. *Image from www.cpslib.org/aip/2000-260.htm, accessed July 28, 2013. Used by permission of the webmaster.*

Fermat dabbled in a wide variety of mathematical fields, but his particular favorite appears to have been what is called the theory of numbers, or number theory. Number theory is a branch of pure mathematics devoted to the study of the properties of the integers, or whole numbers. Number theorists search for patterns in the behavior of the integers and of subsets of the integers—for example, the prime numbers (numbers such as 3, 7, or 23, which are divisible only by 1 and by themselves).

Since a, b, c, and n are all integers in our equation, Fermat's last theorem is an example of a problem from the field of number theory. Fermat himself is recognized as one of the most important figures in the history of number theory. He immersed himself in the study of prime numbers and of "perfect numbers," among other topics. A perfect number is an integer that is the sum of its divisors. The first perfect number is 6, since $1+2+3=6$ and 6 is divisible only by 1, 2, and 3. The next perfect number is 28 ($1+2+4+7+14$). Then comes 496, followed by 8,128. After that, the perfect numbers get really large really fast. To date, only 47 perfect numbers have been discovered, and the largest of them has nearly 26 million digits! An odd fact about the perfect numbers is that all of them discovered to date are even. It has been shown that an odd perfect number, if such a thing exists, would have to be greater than 10^{300}. Proving (or disproving) the existence of an odd perfect number remains an unsolved problem in the realm of number theory.

Of what import is it to understand whether an odd perfect number exists? For that matter, of what import is Fermat's last theorem itself? Not much, as far as anyone can tell. Compared with many of the equations we have discussed in this book, whose implications have revolutionized the way we live, Fermat's last theorem appears to be nothing but a curiosity, albeit an enticing one.

Why do people climb mountains? Because they are there. And so it is with solving mathematical puzzles, for those so inclined. Fermat's last theorem is an interesting proposition, and for over three hundred years after Fermat's tantalizing little off-hand comment, scrawled in the margins of the page, it stood Everest-like, waiting for the mathematical equivalent of Sir Edmund Hillary to come along in the person of Sir Andrew Wiles.

25. Four Eyes

$$\frac{\sin\theta_1}{\sin\theta_2} = \frac{n_2}{n_1}$$

The Snell-Descartes Law of Refraction, or Snell's Law

When light passes from one medium to another, it changes direction. This law describes the relationships between the angles of incidence and refraction when light passes from one medium to another. The indices of refraction for light traveling in two different media are n_1 and n_2. As a beam of light passes through the interface between the two media, the angles that it makes with a plane normal to the interface are θ_1 and θ_2.

No one knows who invented eyeglasses. But there can be little doubt of their historical importance. In an online poll taken at the dawn of the new millennium, distinguished scholars identified eyeglasses as one of the most important inventions of the past 2,000 years. The psychologist Nicholas Humphrey noted that eyeglasses have "effectively doubled the active life of everyone who reads or does fine work, and prevented the world from being ruled by people under forty." The best historical research concludes that the first pair of spectacles was invented by an unknown craftsman near Pisa in 1286. The device consisted of two convex lenses with metal or bone rims, connected by a riveted mechanism allowing the glasses to be clamped to the nose or held in front of the eyes.

Eyeglasses are really just a convenient way to position lenses in front of the eyes, such that the lenses can correct the vision deficiencies of their wearer and thus sharpen his or her vision. The essence of eyeglasses—what really makes them work—is the lens. And while eyeglasses may have been around for more than 800 years, human-made lenses have been around a whole lot longer than that. The oldest such lens so far discovered was fabricated in what is now northern Iraq about 3,000 years ago. The so-called Nimrud lens is about 1.5 inches in diameter and about 1 inch thick. It is a relatively transparent rock crystal that has been ground flat on one side and convex on the other. Researchers speculate it could have been used as a magnifying glass or perhaps even to concentrate sunlight in order to build a fire.

The Nimrud lens did what every lens since that time has done: it bent or refracted light. But lenses were around for perhaps 2,000 years before the physical law that governs their behavior was discovered. What is variously known as the law of refraction, Snell's law, or the Snell-Descartes law was first elucidated by Ibn Sahl of Baghdad in 984 CE, in his book *On Burning Mirrors and Lenses.*

That light bends when it travels from one medium to another has surely been known to humankind for far longer even than the invention of the lens. Early fishermen must have quickly learned that the lovely fish they saw swimming in a clear stream was not located quite where their eyes told them it was, because light rays bend when they pass from air to water. Moving from an observation like that to a mathematical representation of the law of refraction took a long time and represents a prodigious intellectual feat.

Let's start on the right side of our equation above with the ratio n_2/n_1. In this ratio, n is the index of refraction of a given material. The index of refraction is itself a ratio—the ratio of how fast light travels in a vacuum to how fast it travels in a given medium such as water or air. The index of refraction of water is about 1.33, meaning that light waves travel 1.33 times as fast in a vacuum as they do in water. The index of refraction of air is only about 1.0003, meaning that light travels just a little bit faster in a vacuum, such as outer space, than it does in air.

Now it's time to consider the left side of our equation. Do you own a laser pointer? Aim it at the surface of a glass of water. The angle between the laser beam and a line perpendicular to the surface of the water is theta-one (θ_1), as shown in figure 11. And the angle between a line perpendicular to the surface of the water and the laser beam underneath the water is θ_2. These two angles are not equal to each other, since the water bends the laser beam. But by how much? Our equation tells us. If the first angle—the angle the laser makes as it enters the water—is 30 degrees, then the angle under the surface of the water is about 22.1 degrees:

$$\frac{\sin 30^\circ}{\sin 22.1^\circ} = \frac{1.33}{1.0003}.$$

Combining the law of refraction with different-shaped lenses allows you to bend light in all sorts of ways, depending on what you're trying to do.

All of which brings us back to eyeglasses. With different-shaped lenses, eyeglasses can correct for a variety of genetic defects in the shape of the eyeball as well as for the well-known age-related problem with close vision known

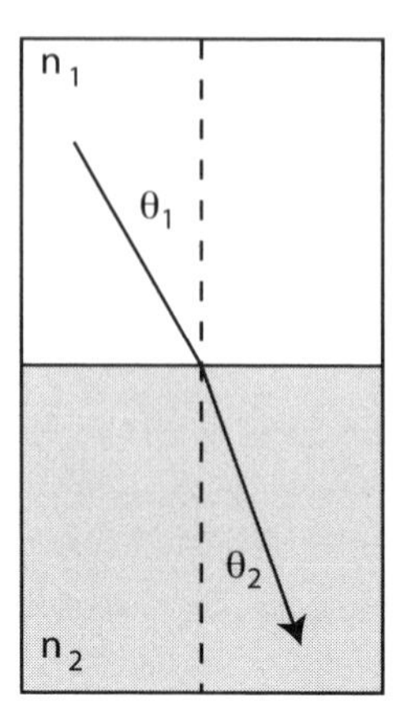

Figure 11. The refraction of a light ray as it passes from one medium to another

as presbyopia. A human eye features two separate lenses of its own. The cornea, on the outside of the eye, has prodigious light-bending power, but its ability to do so is relatively fixed. The eye's other lens, inside the eyeball, is flexible—at least while its owner is young. Muscles inside the eye flatten out this lens when you focus on things far away, and make it round for focusing close up. The whole idea is that the light entering the eye, having passed through both the cornea and the internal lens, is focused on the back of the eye, on the retina, where electronic signals are created and then transferred to the brain.

If the light entering the eye is not precisely focused on the retina, the result is a fuzzy or distorted image. Very often, eyeglasses can help.

It is believed that the first eyeglasses were developed to correct for presbyopia—literally, "aging eye." As the years pile up, the lens inside the eyeball loses its flexibility and can no longer become as round as it did when its owner was young. The result: books, newspapers, computer screens, and the like must be placed farther and farther away from the eye in order to remain in focus. By the time many of us reach our late 40s, the condition has progressed from being an annoyance to something approaching a disability. Appropriate eyeglasses, called reading glasses, can easily correct this problem.

Reading glasses have now been around for over 700 years. Nowadays, eyeglasses can correct for lots of other problems, too, such as near- or farsightedness and astigmatism. All of which gives us older folks a fighting chance to keep the world from being run by those under 40.

26. Bee Sting

$$HN_{SX} = \sum_{j=1}^{J} N_{js} p_{jx}$$

An Equation Modeling the Suitability of a Parcel of Land for a Native Pollinator

Native pollinators require two basic resources: suitable places to nest and a source of flowers. Each is amenable to mathematical modeling. This equation models the suitability, HN_{SX}, of a parcel of land, x, for native pollinator of species s. Parcel x is covered by various types of land cover, $j = 1$ to J. N_{js} is the compatibility of land cover j for nesting by species s. The proportion of parcel x that is covered by land cover j is p_{jx}. This equation is part of a larger mathematical model for predicting the relative abundance of various types of native pollinators in various habitats.

One bite out of every three in the American diet consists of a product pollinated by honeybees. The honeybee is thus an integral part of the $15 billion U.S. produce industry. The list of plants pollinated by bees is incredibly long. It includes almonds, watermelon, strawberries, lemons, limes, apples, cocoa, pumpkins, onions, broccoli, canola, soybean, rapeseed, and cottonseed. The last four plants in that list are used to create, among other things, biodiesel fuel. Pollination, it would appear, provides fuel for other things besides humans and other animals.

All of this is to convince you of the gravity of what has become known as colony collapse disorder. The year 2006 marked the beginning of a devastation of the bee populations in North America and Europe. Beehive owners reported losses of 50% to 90% of their hives. A typical loss rate, after a hard winter, might only be 10% to 25%. In 2006, something different was going on.

Beekeeping is an ancient practice. Evidence of domesticated bee colonies used for honey production dates back four thousand years or more. In ancient Greece, Aristotle wrote extensively about beekeeping. Honeybees are not native to North America. They were first introduced from Europe in the 1600s. Most honeybees in North America live in domesticated hives. There are some feral populations, but these have also declined in recent years.

These days, beekeepers maintain their hives in order to produce honey, wax, or to provide pollination. Pollination services provided by migratory beekeepers date back more than a hundred years in the United States. As a commercial enterprise, pollination by domestic hives is much more common than you might have imagined. Commercial farms have grown so large that there simply are not enough native pollinators—wild bees and other insects—available to service the crops. Modern American agriculture has thus become dependent on migratory beekeeping; what once occurred naturally is now somewhat artificial.

Beekeepers transport their hives from one farming region to the next. Migratory beekeepers thus service the crops that are ready for pollination in one region before moving on to pollinate other crops elsewhere. It's a good business, and as a group, commercial beekeepers earn much more from pollination services than from honey production.

Some in the agriculture community advocate for a return to more natural farming methods, including a reliance on native pollinators. To achieve this it would be necessary for farmers to let some of their lands lay fallow so as to greatly increase the population of native pollinators. Advocates believe native pollination can improve the quality, quantity, and stability of crop pollination compared to a reliance on a single, managed species of pollinator (honeybees).

Our equation above shows one way in which the suitability of land for native pollinators can be assessed. The left side of the equation, HN_{SX}, represents the "suitability" of a given parcel of land for a native pollinator of species s living on a parcel of land x. This is calculated by summing up, on the right-hand side, terms representing the type of land, j (forest, field, lake, and so on); the compatibility, N_{js}, of a given parcel of land, in terms of weather and food supply, for a given type of pollinator; and the proportion of land parcel x that is covered by a given type of land cover, p_{jx}.

The equation is just a mathematical model that allows you to compare different parcels of land with one another and to play "what if" games involving different schemes for managing the land. The bottom line is this: if you want to encourage native pollination, you must provide more natural habitat for native pollinators. It seems counterintuitive, but some studies have concluded that farmers can actually increase overall production from their farms by letting large portions of their lands lay fallow. The increases in natural pollination that are likely to result have the potential to more than compensate for the reductions in the total area devoted to crops.

Not surprisingly, not all farmers believe this, and there are those who are quite happy with the status quo—pollination by managed bee populations controlled by migratory beekeepers. The status quo, however, has been threatened by the collapse in the population of domestic bees with which we began this chapter.

Crashes in the populations of managed bees have occurred before, but the losses due to colony collapse disorder appear to be unprecedented. If the causes of this disaster remain murky, it's not for a lack of effort on the part of investigators. The economic importance of managed bees is such that folks are willing to spend lots of money trying to find out what's killing them. It's likely that there isn't a single cause. One early report claimed that wireless telephone signals were disrupting the bees' ability to navigate and weakening their immune systems. The somewhat less sensational truth is that colony collapse disorder is probably brought on by a combination of factors, including pesticides, parasites, and viruses. The effects of these can be exacerbated by other stresses on managed bee populations, such as hard winters, excessive traveling, and poor nutrition.

Not only is excessive traveling a factor in the overall stress level in bee populations, but it also contributes to the transmission of pathogens among hives. This is a great concern because the decreasing number of domestic hives increases the interaction, nationwide, among the remaining hives. Every year, for example, about 1.3 million hives (roughly half the U.S. population of pollinating beehives) converge in California to pollinate the almond crop. Such large gatherings are breeding grounds for disease.

Honeybees are social creatures—the technical term is "eusocial." Whether in a managed population or in the wild, they cannot live alone, but must depend on their structured life within their hive and colony for their wellbeing. Our understanding of the various factors contributing to that wellbeing is improving but remains incomplete.

The U.S. Agricultural Research Service performs research on the problems described above and tries to keep its eye on the big picture. As of 2012, there have been no unmanageable shortages of domestic bees to pollinate crops. However, our heavy dependency on a single species for such a large portion of our food supply is worrisome.

Our dependence on domestic bees for pollination is emblematic of a larger issue within agriculture: increasing dependence on one or a very few sources for crops—for example, huge factory farms versus a network of small family farms. Single sources provide certain economic advantages but pose the added risk that a problem with that single source can have catastrophic consequences, from widespread outbreaks of *E. coli* bacteria due to contaminated produce to collapses in the population of domestic bees we depend on for pollination. Something to keep in mind the next time you sit down to dinner. Pass the honey, please.

27. Here Comes the Sun

$$SPF = \frac{\text{Time to burn}_{\text{protected}}}{\text{Time to burn}_{\text{unprotected}}}$$

An Equation for Calculating the Sun Protection Factor

The sun protection factor (SPF) is a number used to indicate the effectiveness of sunscreen products. Times to burn, protected and unprotected, are the average values of empirical results from controlled tests on volunteer subjects using artificial sources of ultraviolet radiation.

"Sun is bad for you. Everything our parents said was good is bad. Sun, milk, red meat, college." As Woody Allen reminded us in *Annie Hall*, sunlight really can be bad for you. Get too much of it, and you will regret it, certainly in the short term, but more ominously, perhaps, in the longer term—many years down the road—as well.

Technology, as it often does, comes to our rescue here, in the form of sunscreen products. Miraculous concoctions, these. Rub them on your skin, and they multiply, as our equation above shows, the amount of time you can spend in the sun before burning by a factor known as SPF—the sun protection factor.

All sunscreens have an SPF number printed on the container. Take the amount of time required for you to be burned without protection, say 20 minutes, multiply that by the SPF number, say 10, and you have the amount of time you can spend in the sun without being burned—in this example 200 minutes (3 hours and 20 minutes). Strictly speaking, the equation here is correct only if the radiation you are being exposed to is constant throughout the exposure period. In the real world, this is never the case. Nonetheless, the equation is a useful guide to sunscreen products and to how long they will provide protection.

Sunlight produces radiation across three regions in the electromagnetic spectrum. On the long wavelength end, there is infrared radiation (heat waves). Adjacent to infrared lies the visible spectrum—that part of sunlight that humans can see. The infrared and visible portions of sunlight are not especially damaging to the skin. But on the short wavelength end of the spectrum of sunlight we find ultraviolet (UV) radiation, the stuff that everyone worries about, the radiation that burns your skin, and worse. Only about 8% of the total energy in sunlight is made up of ultraviolet radiation, but that, as it turns out, is more than enough to cause severe damage. The UV region of the sunlight spectrum is often divided into two portions, UVA and UVB. UVB, having shorter wavelengths than UVA, is more energetic and, generally speaking, more dangerous.

SPF ratings are established through controlled laboratory experiments. A group of sun-sensitive people is convened for a two-day experiment. On the first day, a section of each subject's back, unprotected, is exposed to UVB rays (from an artificial source, not the sun). The time it takes for the skin to turn a mild red (the shade of red is quantified) is recorded. On the second day, the same subjects return, and a sunscreen is applied to a different, adjacent location on their backs. The time it takes UVB rays to produce the same mild red color is recorded. The SPF rating is the time it takes to burn with the sunscreen divided by the time it takes to burn without sunscreen. Since everyone burns at different rates, the SPF ratings for the group are averaged and rounded down.

The chemists who concoct sunscreens have gotten better and better at their jobs, and over the years SPF values have continued to increase. The lowest possible SPF number is 2; as of this writing, the highest values are up around 130. Some question whether high-SPF sunscreens are a better solution, or whether it is safer simply to reapply a low-SPF sunscreen more frequently. The SPF testing described above involves only UVB, and sunscreens tested this way don't necessarily protect against UVA. To find a sunscreen that will protect you from both UVA and UVB, look for labels that advertise "broad spectrum" protection.

A higher SPF number indicates longer possible protection against burning, but no sunscreen, no matter the SPF, can provide 100% protection from UVB rays. At an SPF of 15, about 94% of UVB rays are absorbed by the sunscreen. Double the SPF to 30 and you take care of 97% of the UVB. At an SPF of 50, about 98% of UVB rays are absorbed. The Food and Drug Administration, which oversees SPF ratings, has been considering putting a cap, or maximum value, on SPF so that consumers won't misinterpret high SPF numbers as a guarantee of all-day, 100% protection from the sun.

There is a difference between sunscreen and sunblock. Sunblock is made from compounds like zinc oxide or titanium oxide, which form a barrier layer and reflect the sun's harmful rays. This is the white stuff folks sometimes apply to their noses. Sunscreens, on the other hand, do more than simply shield your skin from harmful UV rays. Sunscreen is a combination of inorganic and organic ingredients that either block or absorb UV radiation. Sunscreens often do contain small particles of zinc oxide to reflect the sun's rays, but the particles are typically too small to be seen. The organic ingredients in sunscreen protect your skin by absorbing the harmful wavelengths of light, much as water molecules absorb microwave radiation (and thus get warmer) inside a microwave oven. Since the molecules in sunscreen absorb the harmful radiation, that radiation is not able to interact with, and damage, your skin.

Choosing a sunscreen these days can be daunting. High SPF ratings do provide longer protection, but the average person applies only about half the amount of sunscreen that is used in the laboratory tests. If this describes what you do, you are getting only about half the protection indicated by the SPF label on the bottle. And if you swim, sweat, or wipe your skin, you will need to reapply the sunscreen more often. Some sunscreens claim to be water- or sweat-resistant, but let the buyer beware. For the best protection from your sunscreen, it is recommended that you apply it 30 minutes before being exposed and then again 30 minutes after first exposure to the sun. Then reapply every two hours after that.

Your mother told you to go outside and play. It was good advice then, and it still is. Just don't forget the sunscreen.

28. A Leg to Stand On

$$F = \frac{\pi^2 EI}{(KL)^2}$$

The Euler Buckling Equation

The Euler buckling equation predicts the force necessary to cause a long, slender column of a given geometry to fail by buckling. The compressive force, F, required to cause a column to buckle is a function of the column's length, L; the elastic modulus, E, of the (homogeneous) material of which the column is constructed; and the area moment of inertia, I, of the column's cross section. The constant K depends on the type of supports at the ends of the column.

Have you ever wondered why an ant wears its skeleton on the outside, while yours is internal? The exoskeleton works quite well for the ant, so why not for you? Part of the answer can be found in the Euler buckling equation. "Buckling" is a technical term that refers to one specific way in which structures can fail. Buckling is very easy to demonstrate. Hold a plastic soda straw with one index finger on each end of the straw. Now press your fingers towards each other, compressing the straw. After a few pounds of force applied this way, the straw begins to bow, suddenly develops a kink—and buckles.

Buckling is an insidious failure mechanism, since it can cause structures (such as bridges or buildings) to collapse, suddenly and without warning, under loads that are only a small fraction of the load that the material in question is capable of carrying. Leonhard Euler (1707–83), the incomparable Swiss mathematician and physicist, published the equation above in 1744. The force, F, required to buckle a long, slender column (such as our soda straw) is proportional to E, the elastic modulus (or stiffness) of the material, and I, the area moment of inertia of the cross section of the column. In the denominator of the equation, K is a constant that depends on how the column is supported at its ends and L is the length of the column.

Notice that the L in the denominator is squared. Cut a soda straw in half with a pair of scissors and try to buckle it with your fingers as before. Since the length, L, has been cut in half, Euler's formula predicts that the buckling force, F, will be four times as great. Likewise, if you had a soda straw twice as long as a regular one, you could buckle it with one-fourth the force required for the regular straw. Our soda straws of varying lengths just barely begin to provide us with an explanation for why an exoskeleton works well for an ant and not so well for a person.

The exoskeleton is quite popular in the insect world but certainly not limited to it. In addition to structurally supporting the body of its possessor, an exoskeleton also provides armorlike protection—a feature not offered by the endoskeletons of humans, other mammals, birds, and so on. Some animals, such as the tortoise, possess both endo and exoskeletons. The armadillo is a mammalian example of an animal with an exoskeleton of sorts. Crustaceans such as crabs and lobsters possess exoskeletons. Hermit crabs scavenge for abandoned seashells, adopting them as their exoskeleton. They are thus obliged to search for new shells as they outgrow their old ones. This is true of creatures with natural exoskeletons as well. An evolutionary disadvantage to exoskeletons, both natural and adopted, is thus that they allow for only limited growth. Insects, crabs, and other creatures with exoskeletons are sometimes obliged to shed their exoskeletons as their bodies grow, leaving themselves, for a time, extraordinarily vulnerable to predators.

Back in the 1950s and '60s, scary science fiction movies sometimes featured creatures like ants, spiders, and cockroaches that had grown as big as houses (often as a result of a nuclear experiment gone wrong). These films weren't really all that scary, and they certainly weren't all that realistic. Creatures that large with exoskeletons create some unique structural challenges. Euler's buckling equation helps explain these.

For example, imagine a creature whose body, a good-sized potato, is supported by six soda-straw legs stuck in the bottom of the potato. What would happen to the buckling resistance of those legs if the creature were 10 times as large? The soda straw legs would then be 10 times as long with 10 times the diameter and 10 times the thickness. If you calculate the critical buckling load using Euler's equation, you will find that the larger straws (made of the same material as the smaller ones) would have a critical buckling load 100 times as great. This sounds promising: the straw is only 10 times larger, and yet it will hold 100 times as much weight without buckling.

There's just one problem. When the potato body of our creature is made 10 times as large, its weight increases, not 10 times or 100 times, but 1,000 times! Imagine a roughly cylindrical potato 3 inches in diameter by 6 inches long. Make it 10 times as large, and now it's 30 inches in diameter by 60 inches (5 feet) long. That's one heck of a potato. And it will weigh *1,000 times* as much as the smaller potato. A one-pound potato becomes a 1,000-pound monster when you make it "10 times" as large.

Each leg on the larger creature is thus required to support 1,000 times as much weight as the leg on the smaller creature must hold up. So even though the larger leg, according to Euler, will hold up 100 times as much weight without buckling, it's in big trouble, because it is being asked to hold up 1,000 times as much weight.

The larger creature in the above example would be lucky to be able even to support its own weight without buckling its legs. Carrying the loads involved in moving around, jumping, and so on would be even more perilous. Scaling up a creature with an endoskeleton is no picnic either, but that endoskeleton creature appears to have certain evolutionary advantages over an exoskeleton one with respect to load-carrying ability; buckling resistance is only one of those advantages.

The subject of "scaling" in biology is interesting and complex enough that it has become its own subspecialty. The implications of making living creatures much larger (or much smaller) influence all kinds of things. Buckling resistance is one reason that an ant the size of an SUV wouldn't work, but it's far from the only one. Many have pointed out, for example, that the way insects breathe is far less amenable to scaling up than the way mammals breathe. Mammals deliver oxygen to tissue by means of blood, a liquid, which picks up the oxygen in the lungs. Insects don't have blood and thus deliver oxygen directly (as a gas) through a series of tubes, called tracheae, crisscrossing their bodies. Tracheae carrying gaseous oxygen don't scale up as well as blood vessels containing dissolved oxygen do. As an insect's body gets larger, the amount of space that must be devoted to tracheae goes up rapidly, effectively limiting the size of insects.

Which would happen first to a giant ant: would it suffocate, or would it collapse under its own weight? It's hard to know. In any event, it probably wouldn't be around long enough to star in its own movie.

29. Love Is a Roller Coaster

$$emf = -\frac{\Delta\phi_B}{\Delta t}$$

One Form of Faraday's Law of Induction

The equation quantifies the electromotive force, *emf*, that is generated by the rate at which a magnetic field, ϕ_B, is changing with respect to time, t. The minus sign signifies, in accordance with Lenz's law, that the induced electromotive force and the change in magnetic flux have opposite signs.

Folks sure are passionate about their roller coasters and other amusement park rides. They'll stand in line for hours to ride them. Part of the thrill of these rides comes from the sensation of motion as you hurtle down the track through impossible twists, turns, and loops. But let's be honest: part of the thrill comes from the danger involved. Amusement park rides have a pretty good safety record, but it's not perfect, a fact that has to be in the back of your mind as they strap you in to that padded steel cage.

One thing that helps keep you safe on many of those rides is a braking system with no moving parts. The braking systems on almost all wheeled vehicles—cars, bicycles, and even airplanes (on the ground)—are based on friction. When you step on the brake pedal in your car, you cause a component that is rotating with the wheel to come into contact with other components that aren't rotating. The resulting friction transforms the kinetic energy (the energy of motion) of the car into thermal energy (heat). And so you slow down.

Friction-based braking systems have been under development for hundreds of years, and they work pretty darn well. And they have plenty of applications on amusement park rides such as roller coasters. But because these rides travel on tracks, they are especially amenable to another braking system based on a completely different concept: that of magnetic induction. In amusement parks, magnetic induction braking is most commonly used on so-called "free-fall" rides. On such rides, the track structure contains built-in copper plates. The freely falling car or cabin contains (along with you and your friends) a set of powerful permanent magnets. As the passenger car falls, the magnets pass by very close to the copper plates. This induces a powerful force that opposes the motion of the car, and the car slows down.

This effect can be very easily demonstrated. (An early demonstration is attributed to the French physicist François Arago [1786–1853].) One way to do it is with a straight piece of copper tubing about a foot long, a solid cylindrical magnet that just fits inside the tube, and a cylindrical piece of steel the same size as the magnet. Hold the copper tube vertical, and drop the magnet into it. Repeat with the piece of steel. You should notice that the steel descends through the tube much more rapidly than the magnet.

The more powerful the magnet you use and the closer its diameter is to the inside diameter of the copper tube, the more convincing your demonstration will be. That is, the greater the braking effect will be on the magnet as it drops through the tube. You can see why this is such an attractive (no pun intended) braking system. The effect is automatic, it doesn't require any external power source, and it has no moving parts. It's just about as foolproof as anything could be.

Copper, of course, is a nonmagnetic material—it is not attracted to the magnet the way a steel plate would be. The braking effect on the magnet in this demonstration, or on the free-fall amusement park ride, is due to the forces induced by the magnet as it moves past an electrically conductive material (the copper). These forces are predicted by Faraday's law of induction and by Lenz's law. Michael Faraday (1791–1857) was a brilliant and prolific British physicist and chemist whose long list of discoveries and inventions continues to influence modern science and technology. Heinrich Lenz (1804–65) was a Russian physicist.

In Faraday's law of induction, one of the many forms of which is shown in our equation above, *emf* is the electromotive force that is generated, in this case, by the rate at which a magnetic field, ϕ_B, is changing as a function of time, t: $\Delta\phi_B/\Delta t$. Faraday's law thus predicts the magnitude of those electromotive forces, while Lenz's law determines the direction of the forces. The minus sign on the right-hand side of the equation comes from Lenz's law, and it tells us that in the case of the magnet in the tube, the force generated by the magnet as it drops through the tube *opposes* the motion of the magnet. In other words, that it is a braking force.

Very similar physical phenomena are at work in lots of other modern technological devices. For example, do you drive a Toyota Prius or some other hybrid-electric vehicle? Or maybe an all-electric car such as a Tesla Roadster? Electric vehicles and gas-electric hybrid vehicles nearly always have two braking systems that work in tandem: a conventional set of hydraulic friction brakes along with a "regenerative braking" system. Let's consider only the regenerative brakes. The electric motor in a vehicle like that will drive the wheels using battery power. But when you step on the brakes, the motor acts as a generator. It converts the kinetic energy in the motor's rotating shaft into electrical energy, in the process slowing down the shaft (and thus braking the car). The electrical energy created in this process can be stored in the car's batteries. In friction braking, 100% of the kinetic energy in the vehicle is lost during braking (you simply convert the energy into heat). With regenerative braking, half of that energy, or more, can be recaptured—regenerated—and stored in the car's batteries. Depending on the type of electric motor used, the physical principle involved is very similar to that in the two examples above: the magnet in the copper tube or the amusement park free-fall ride.

Magnetic induction, as noted above, is routinely used to slow down amusement park rides, but it can also be used to *start* such rides. Traditionally, chain drives have been used in roller coasters to propel the vehicles uphill, whereupon gravity does the rest. In a relatively recent trend, the vehicles are "launched" down the track at the beginning of the ride. Such launches feature high accelerations, contributing to the thrill (at the expense of the mounting suspense engendered by the old chain drives, as you lurch ever so slowly up that impossibly steep first hill). Launched roller coasters can employ a variety of technologies to supply the launching energy, only one of which is a linear induction motor (LIM) using the physical concepts discussed above.

Michael Faraday never got to ride a roller coaster, but he might have been imagining them when he remarked, "Nothing is too wonderful to be true if it be consistent with the laws of nature."

30. Loss Factor

$$\tan\delta = \frac{E_2}{E_1}$$

The Loss Factor, Tan δ, for a Viscoelastic Material

E_1 is the storage modulus for the material, and E_2 is the loss modulus. The angle δ relates to the mismatch in the time scales between the forces acting on a viscoelastic material and the resulting deformation of that material. As the force applied to a material and the resulting deformation get more and more out of phase with each other, the angle δ increases, and so does tan δ. For a given material tan δ changes with temperature. The temperature at which tan δ is a maximum is called the glass transition temperature.

Pity the poor O-ring, that most humble of components. When it does its job properly, no one even knows it's there, aside from the engineer who specified it and the technician who installed it. When it doesn't, stuff leaks out that shouldn't. Often, the consequences don't go much beyond annoyance, irritation, and inconvenience. But on January 28, 1986, things were different. Spectacularly, catastrophically so. For that was the day that the space shuttle *Challenger* went up and then, a little more than a minute later, disintegrated in midair, ending the lives of its seven crew members.

It is unfair, and incorrect, to blame the *Challenger* disaster entirely on the infamous O-ring seals in the shuttle's solid rocket boosters (SRBs). Disaster experts sometimes speak of a "failure chain" that is often present when a catastrophe like the *Challenger* accident occurs. A failure chain is a sequence of events or conditions that come together, against all odds, and result in tragedy. These might include a defective component, a malfunctioning instrument, a human error, an unusual meteorological condition, a failure to communicate (or a miscommunication), and so on. When they all occur simultaneously (or sequentially), the result can be disaster. Had any one of those events or conditions been missing, the calamity would have been averted.

The O-ring seals on the *Challenger*'s SRBs were just one of the items in the *Challenger*'s failure chain. Others included the design of the joints in the SRBs themselves (the joints the O-rings were supposed to seal), the temperature at launch (30°F, the coldest shuttle launch ever, by far), a series of wind shear events (the worst ever experienced during a shuttle launch) about 30 seconds into the flight, and poor communications between NASA and some of its various contractors.

But our story here is about the O-rings themselves, or more precisely about how the rubber that they were made of behaves at low temperatures. Rubber is an elastomer, which is a type of polymer. Elastomers are capable of undergoing large amounts of elastic (or reversible) deformation. Stretch a rubber band between your fingers. You can stretch it a great deal, and yet when you release it, the rubber instantly returns to its original shape. This behavior is unique in the world of materials. If you repeat our little rubber band experiment at low temperatures, however, the results are strikingly different.

Take a rubber band and stretch it over a block of wood. Immerse the wood block and rubber band in liquid nitrogen for a few seconds, remove them from the liquid, and then carefully pry the rubber band off the wood. At first, the rubber will maintain, as if by magic, its stretched-out shape. As the rubber warms up, it gradually shrinks back to its original shape. It all makes for a great classroom demonstration of something called the glass transition temperature. Liquid nitrogen boils at −196°C (−321°F). At that temperature, the rubber in a rubber band is in a "glassy" state, well below its glass transition temperature. At room temperature, well above the glass transition, rubber is, well, rubbery—very stretchy. When rubber is in the glassy state, it is extremely rigid, as in our little classroom demo.

As the temperature goes down, the transition from rubbery behavior to glassy behavior is relatively gradual. You can show this with another great classroom demo. Get a couple of small rubber balls. At room temperature, a rubber ball is rubbery (no surprise there), and it bounces well. Chill it in liquid nitrogen and bounce it again. (Be careful: at that temperature it is extremely brittle and will crack like a piece of fine china.) At cryogenic temperatures, a rubber ball, when bounced, gives off a distinctive glassy ring; it sounds like you're bouncing a marble. So far, this demo isn't much different from our earlier one with the rubber band and the block of wood, but stay tuned. Now put a rubber ball in your freezer (typically about 0°F (−18°C)) for at least 30 minutes. When you take it out and bounce it, you get neither rubbery (normal) nor glassy (sounds like a marble) behavior. Instead, the ball hits the ground with a thud, yielding an anemic rebound that is only a small fraction of the height of its rubbery room-temperature bounce. The bounce is so pathetic it almost seems as if the ball were made of leather. In fact, that's what it's called—leathery behavior.

Our equation here shows one way that the glass transition temperature—where an elastomer's behavior is at its most leathery—is determined. A quantity called tan δ (pronounced "tan delta" and also called the loss factor) equals the ratio of E_2 to E_1.* E_2 and E_1 are called the loss and storage moduli for the material in question. To roughly approximate what this all means, get out your rubber ball again. At room temperature, when you drop it on concrete from five feet, it might bounce three feet. As you cool it down from there, its bounce gets progressively lower. As it loses the ability to bounce, we say its loss factor (tan δ) is increasing. Eventually, it gets to a temperature at which its bounce is the absolute lowest. That is the glass transition temperature. Below that temperature, the material becomes more glassy and bounces (higher) like a marble. Above that temperature, it becomes more rubbery and bounces higher the way a rubber ball should. Right at the glass transition temperature, we have minimum bounce and thus maximum leathery behavior.

Which brings us back to the SRB O-rings in the *Challenger.* The temperature at launch, about 30°F (−1°C), rendered the SRB O-rings leathery in the extreme. At that temperature, the rubber in the O-rings was sluggish and slow to respond to applied stresses. Just as a racquetball removed from the freezer is sluggish and slow to respond to the stresses induced in bouncing it, the O-rings could not respond quickly enough as the joint they were supposed to seal twisted and deformed under the enormous stresses involved in a shuttle takeoff, stresses that were then compounded by the vicious wind shear events about 30 seconds later. The joints the O-rings were supposed to seal thus leaked hot combustion gases, eventually leading to disastrous consequences.

The *Challenger* disaster gripped the nation's consciousness. As with the assassination of John F. Kennedy 23 years before, practically every American old enough to remember it could tell you where he or she was when they heard the news of the *Challenger.* In its wake, the U.S. government established the Rogers Commission to investigate every aspect of the disaster. And without

*The angle δ in tan δ relates to the mismatch in the time scales between the forces acting on a viscoelastic material (our rubber ball) and the deformation of the material. As force and deformation get more and more out of phase with each other, the angle δ increases, and so does tan δ.

question, the single most famous moment in the Rogers Commission inquiry involved the SRB O-rings.

Not coincidentally, that moment also involved the most celebrated member of the Rogers Commission: the brilliant, irascible, and incurably curious Nobel Prize–winning physicist Richard Feynman. Feynman took a small rubber O-ring and pinched it in the middle with a little steel clamp. He then placed the clamped O-ring in a glass of ice water. With the TV cameras running, Feynman pulled his experiment from the water with a flourish and removed the clamp. Much like our earlier wood block and rubber band test, Feynman's O-ring did not immediately snap back to its original "O" shape. Having been chilled below its glass transition, it needed lots more time to react. It was a smoking-gun moment, and it left many Americans with this conclusion: the *Challenger* crashed because its O-rings got too cold. That, as we have seen, is the truth, but not the whole truth.

31. A Slippery Slope

$$F_f = \mu N$$

Amontons' First Law of Friction

In order to move one body that is in contact with a second body, friction must be overcome. The force due to friction is directly proportional to the applied load between the two bodies. In the equation, F_f, the force necessary to overcome static friction between the two bodies, is equal to the static coefficient of friction, μ, between the two bodies times N, the normal force of one body on the other.

During a crucial stage of a competition, an athlete slipped and fell, costing his team the victory. In the press conference after the game, he described himself as having had a "mu times *N* problem." Now that's our kind of sports hero.

Friction keeps athletes from falling down (whereas its lack causes them to fall). Friction also stops your car when you hit the brakes. Not only that: it causes your car to move when you hit the gas; without it, your tires would slip, as they do on ice. Whenever one solid moves relative to another that it is in contact with, friction is involved. Friction is not limited to solids, however. When an airplane flies, the air passing over the plane slows the plane down. This so-called aerodynamic drag force is a type of friction. The same force acts on moving cars, and water exerts a similar force on moving boats.

Sometimes referred to as the "friction equation," sometimes as Amontons' first law of friction, and sometimes as the Amontons-Coulomb law, $F_f = \mu N$ states that the force of "static" friction acting on one body resting on another body is equal to the coefficient of friction times the normal force, N. Here's an example: A wooden crate weighing 500 pounds is resting on a flat concrete surface, the floor of your garage. A rope is tied to one side of the crate. How much horizontal force would you need to exert on the rope to get the box to begin to slide across the floor? To make the box move in this way, you have to overcome static friction, which our equation tells us is equal to μN. N is just the weight of the box, 500 pounds. The friction coefficient μ in this example is close to 0.6. Thus, μN is equal to 0.6×500, or 300 pounds. How much force would it take to begin sliding the same box, in the same manner, across the frozen surface of an ice skating rink? About 25 pounds, since the coefficient of friction between ice and wood is about 0.05.

The history of this important little equation is a little bit murky. Guillaume Amontons (1663–1705) was a French scientist who made important contributions in several areas, including the thermodynamics of gases, in addition to his studies on friction. He was an accomplished instrument maker who improved such devices as the barometer and the thermometer. As a scientist, he is sometimes given credit for the concept of "absolute zero" in temperature—something that was later quantified by Lord Kelvin (and then named after him). But Amontons is probably best known for friction.

Friction is categorized as part of a subset of physics called tribology. It is now believed that Amontons' main discoveries related to friction, including the first law of friction, were explained earlier by none other than Leonardo Da Vinci. Leonardo's tribological discoveries were recorded in journals that were not discovered until well after Amontons' death. Later, Charles Augustin de Coulomb (1736–1806) came along and both verified and expanded upon Amontons' work. Bracketed by giants—Leonardo before, and Coulomb after—it's not hard to see why Amontons tends to get lost in the shuffle. He didn't publish much, seeming to prefer to toil in his laboratory and let his work speak for itself. And he was also deaf, which may have inhibited his ability to make his work known to others. Most of the little that is known of Amontons' life comes from a brief biography by Fontenelle, written upon Amontons' death and published in 1705 as part of an annual history of the French Academy of Sciences, of which Amontons was a member.

Amontons also discovered what is sometimes called Amontons' second law of friction, which states that the force of friction between two bodies is independent of the area of contact between them. Let's go back to our earlier example of the wooden crate on the concrete floor of a garage. It really doesn't matter how large the crate is. All that matters is how much the crate weighs. Two 250-pound crates, stacked on top of each other, would require a horizontal force of 300 pounds on the rope to make them move (per our example above). That force would not change if the crates were sitting side by side on the floor and being dragged sideways by a single rope.

This second law is really implicit in the first, since in the first law the friction force depends only on μ and N, and not on the surface area (or anything else). This is a counterintuitive result, and there are exceptions, but it's true most of the time, and it certainly makes things easier for anyone trying to calculate friction forces.

Coefficients of friction (μ in our equation) have been measured and tabulated for all kinds of things. Wood on concrete, as in our earlier example, is about 0.6. Steel on steel has a high coefficient, 0.8. Machines with steel parts that are in contact and move relative to one another thus require lubrication to function properly. At the other end of the spectrum, Teflon on Teflon has a low coefficient of 0.04. Everyone knows that Teflon is slippery; its coefficient of friction quantifies it for us.

If you look closely at tables of friction coefficients, you will notice that they often include not one but two coefficients for a given pair of materials. One is the static coefficient that we've been discussing, while the other is the sliding (or kinetic) coefficient. Think about a rubber tire on a concrete road. When you hit the brakes, the brake calipers pinch down on the brake's disks, creating friction that slows the wheel. But your ability to stop depends not only on the friction between brake caliper and disk but also, critically, on the friction between the tire and the road. Under normal braking conditions, the wheel slows down, but continues rolling on the road. Thus, the contact between the road and tire is "static" and not "sliding." Sliding contact is what occurs when you slam on the brakes and lock the wheel, such that it skids across the road. The coefficient of sliding friction is nearly always less than that of static friction, and thus your car slows down more quickly when the wheel is still rolling than if it is locked up and sliding. This is one reason why most new cars nowadays feature "antilock" brakes—brakes whose built-in computer controls prevent the wheels from locking up and sliding. The other big advantage of antilock brakes is handling. A locked-up wheel can't be steered. So, as long as your wheels are rolling, you can stop faster, and you are less likely to lose control of the car.

If you slip and fall during an athletic event, blame it on friction, or the lack thereof. But if you don't fall and manage to score the winning goal, just remember that you have friction to thank for your good fortune.

32. Transformers

$$f(x) = a_0 + a_1\cos(x) + a_2\cos(2x) + a_3\cos(3x) + \cdots$$
$$+ b_1\sin(x) + b_2\sin(2x) + b_3\sin(3x) + \cdots$$

A Fourier Series

In this Fourier series a periodic function, $f(x)$, is represented as the sum of a series of sine and cosine terms. The function $f(x)$ must be periodic and be "reasonably behaved." If so, it can be expressed as the sum of a series of sines and cosines. The coefficients a_n and b_n can be determined from $f(x)$ using formulas attributed to Euler. The more terms that are used on the right side of the equation, the closer their sum approximates $f(x)$. A graphic example is shown in figure 12.

You would be hard pressed to find a person in the developed world whose life is not influenced, every day, by the math of Joseph Fourier. Do you have a cell phone, an iPad or other tablet device, a digital camera, or a personal computer? We thought so. Each of these staples of modern life utilizes the transformative mathematical techniques developed by Fourier: the Fourier series (as exemplified by our equation above) and its cousin, the Fourier transform.

Jean-Baptiste Joseph Fourier (1768–1830) was a French mathematician and physicist. The influences of his work on modern society are much greater than those of any number of scientists whose names are far better known. Scientists and engineers do know the Fourier name, however, through the famous Fourier transform. Fourier's great contribution to modern technology was his discovery of the mathematical means by which really complicated signals (such as sound, radio, or light waves) can be mathematically broken down (transformed) into bite-sized pieces, manipulated in some way, and then retransformed into a new, improved version of the original signal. Everyone in the developed world is exposed, every day, to countless examples of signals that have been manipulated by Fourier's mathematics. Our equation here is a Fourier series, in which a periodic function, $f(x)$, is approximated by a sum of sine and cosine terms.

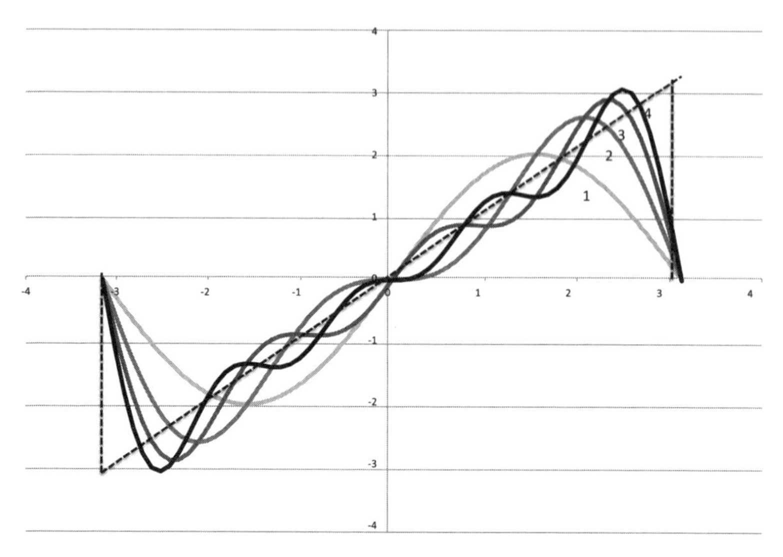

Figure 12. Graphic example of a Fourier transform. The function being approximated is a "saw tooth," represented by the dotted line. The four curves shown, numbered 1 through 4 on the graph, are the first four partial sums of the Fourier transform for the saw tooth. As the number of the curve increases (as more and more terms are included in the Fourier transform), the curve more closely approximates the saw tooth.

A good modern-day example of the application of Fourier's math is the music we store on devices such as the iPod. An iPod is a type of MP3 player, since the music files it stores and plays back for us are encoded in the format known as MP3. The MP3 format has been around since 1992. It was developed as a means of storing large amounts of digitized music in a relatively small amount of electronic storage space, such as on a computer's hard drive. Before MP3, about 10 megabytes of storage were required to store one minute's worth of high-quality music. The MP3 format cuts this by a factor of about 10, allowing one minute of music to be stored in about 1 megabyte.

In a nutshell, here's how this works. When we listen to music—say, a live orchestral performance—a significant portion of the sound waves being created by all those instruments is essentially inaudible to the human ear. Yet when that same music is captured electronically, all the information it contains—what we can hear along with what we can't—is included. The algorithm that results in an MP3 file first chops up the music mathematically into hundreds of individual frequency bands. Within each band, information inaudible to humans is then removed. Finally, all the bands are put back together, and you get the music back in MP3 format. Except now, with all that unnecessary information removed, it can be stored electronically in far less space, making devices like the iPod practical. And Fourier math is at the heart of it all. The great Lord Kelvin once said, "Fourier is a mathematical poem." He couldn't have known how close to the truth he actually was.

Applications of Fourier's math go far, far beyond the MP3 example above. Since we've been discussing music, let's stay there for a moment. Music and math are as tightly intertwined as the DNA double helix (see, for example, our discussion of Pythagoras in chapter 23). And there is perhaps no better place to appreciate the connections between music and math than in the math of Joseph Fourier. Two musical instruments each playing the same note (a piano and a guitar, for example) will sound very different. This is because musical instruments do not produce "pure" tones but, rather, compound tones that are rich in overtones. Overtones are what give various instruments their characteristic "color." The sound produced by a musical instrument is a combination of the fundamental frequency of the note being played and the overtones. Fourier's math allows us both to *synthesize* and to *analyze* complex musical tones. We synthesize a tone by starting with a pure tone and adding to it, one by one, various overtones—to give the note its color. We analyze a compound tone by breaking it down into its component pure tone and overtones. The human ear is sensitive enough to analyze a compound musical tone into its pure tones and harmonics and thus to hear them all separately, in a fashion analogous to a glass prism that splits sunlight into a rainbow of colors. In this respect, the ear is performing Fourier analysis.*

Joseph Fourier developed the math that we now use to synthesize and analyze music in order to understand the ways in which heat flows. His *Analytical Theory of Heat*, a seminal work, appeared in 1822. In attempting to solve the complex equations that describe the flow of heat along a metal rod, for example, Fourier introduced techniques applicable to the mathematical modeling of an astonishing range of phenomena, from music, to heat flow, to vibration, and on and on and on.

Fourier is also often given credit as being one of the first to discuss, and to attempt to mathematically model, what is now known as the planetary

*The ear is capable of distinguishing the various overtones that go into a compound tone. In the analogous situation, however, the unaided human eye is incapable of discerning the various individual colors that have been mixed together to form other colors, as, for example, when blue and yellow light are mixed to create green light.

"greenhouse effect." In 1824, Fourier published a paper discussing the temperature of the planet Earth and the atmosphere. Without identifying any greenhouse gases by name, Fourier nonetheless described aspects of the planetary greenhouse effect: sunlight arriving from space passes easily through the atmosphere and is then absorbed by, thus transferring its energy to, the oceans and dry land. The water and land warm up as a result. The warm water and land, in turn, give off heat, not as visible light but in the form of infrared radiation, which Fourier referred to in the language of his day as "obscure heat." Fourier realized that it was much more difficult for infrared radiation to pass back out of the atmosphere than it had been for the same energy to enter the atmosphere, earlier, in the form of visible sunlight, and he thus concluded that this must be why the planet stays warm. The first scientist credited with identifying individual greenhouse gases and attempting to quantify their effects, the Swede Svante Arrhenius in 1896, acknowledged the pioneering role of Fourier in this field.

In his work on planetary heat, Fourier realized that the overall temperature of the planet could be calculated by summing up the individual contributions of all the various energy sources at work here: incoming sunlight, outgoing infrared radiation, the conduction of heat outwards from the Earth's molten core, and so on. In other words, these effects could be considered independently, one by one, and then added up to describe the overall phenomenon. This describes a crucial aspect of Fourier's scientific philosophy, but even more so, of his technique of chopping up a complicated mathematical function into manageable pieces, manipulating those pieces, and then reconstructing the original function in a new and improved form.

Fourier's place among the greatest and most influential scientists of all time is thus secure. As a young man, however, Fourier narrowly escaped being chopped up into pieces himself, in the guillotine of the French Revolution. Had that indeed been his fate, it is rather doubtful that his own mathematics, however elegant, could have pieced him back together again.

33. A House of Cards

$$a_n = ar^{n-1}$$

One Example of a Geometric Progression

The *n*th term in this progression, a_n, is calculated by multiplying a, the first term, by r^{n-1}, where r is the ratio between successive terms (often referred to as the "common ratio"). The geometric progression expressed in the equation above may also be written as $a_n = ra_{n-1}$. From this latter relation, it is clear that r is the ratio between successive terms in the progression: $r = a_n/a_{n-1}$.

"Charlie, you're the greatest Italian who ever lived!" Or so someone supposedly remarked to Charles Ponzi at the height of his brief but dizzying fame in Boston in 1920. Ponzi allegedly demurred, citing instead his compatriots the explorer Christopher Columbus and Enrico Marconi, the inventor of radio. Marconi may have invented radio, came the reply, "But you, Charlie, you invented money!"

And so, for a time, it appeared that he had. Or he appeared somehow to have invented a way to make money appear from out of thin air. The sort of illegal financial scheme that immortalized Ponzi's name was first put into practice much earlier, however. Ponzi was just better at it than almost anyone else, at least until a guy named Bernie Madoff came along.

A Ponzi scheme is a form of financial fraud. The basic idea is simple. Someone convinces a few people to invest in a seemingly legitimate money-making strategy. (We'll get back to Ponzi's own personal example in a moment.) Investors buy in, lured by promises of lucrative returns. But the supposed investment vehicle doesn't really exist. Those running the scheme simply pay off the early investors with proceeds from later ones. Word gets around—the early investors are making a killing—and soon everyone wants a piece of the action. But there's a catch. Each new generation of investors has to be much larger than the one before it in order to make the necessary payoffs, and eventually the whole thing collapses under its own weight.

For example, imagine that a Ponzi schemer has two original investors (the first generation). To pay them off, let's say he needs four new investors (the second generation). To pay those folks will require eight more. And then 16, 32, 64, and so on. The series 1, 2, 4, 8, 16, 32, 64, . . . is an example of a geometric progression, and it is represented by our equation above. The equation calculates a_n, the *n*th term in the progression, if *a* is the original value (1 in this example) and *r* is the ratio between successive terms (2 in this case). The seventh term, a_7, in the progression 1, 2, 4, 8, . . . is thus equal to $1 \times 2^{n-1} = 2^6 = 64$. But this also means that the 20th generation of investors would include 2^{19}, or 524,288 individuals! By its very nature, a true Ponzi scheme is doomed to failure from the moment it starts.

In 1920, Charles Ponzi was an Italian immigrant with a checkered past living in Boston. A letter he received from Spain gave him an idea. The letter contained something called an international reply coupon, which had been purchased in Spain but which the recipient (in the United States in this case) could redeem for the postage for a return mailing to Spain. Because of differences in exchange rates, it was theoretically possible to make a rather large percentage of profit by buying international reply coupons in one country and redeeming them in another. Ponzi decided to go into business doing so. There was nothing illegal about this, and as noted above, it was potentially quite profitable. But the amount of money to be made on a single coupon was very small, and redeeming the coupons required lots of time and effort because of the red tape involved.

Ponzi soon realized that what was theoretically profitable was practically unworkable. But by then he'd enticed a number of folks to invest in this business—and he'd promised them huge returns on those investments. He quickly abandoned the impractical reply coupon scheme and set about paying off earlier investors with the money he received from later ones. Ponzi's scheme quickly went viral, to borrow a modern term. In just a few months, he had made hundreds of thousands of dollars—and a dollar in 1920 was worth more than 10 times as much as a 2012 dollar. Ponzi's meteoric rise drew suspicion almost immediately. After all, he was promising to double investors' money in just a few months. But when a Boston writer suggested something was fishy, Ponzi sued him for libel and won $500,000.

By early August 1920, however, less than a year after it had begun, Ponzi's financial house of cards collapsed. He spent years in jail and was eventually deported. Ponzi died in Brazil, in 1949. Near the end, he finally admitted that the Boston adventure that had immortalized his name was nothing but a swindle.

Bernie Madoff, born in 1938, founded Bernard L. Madoff Investment Securities in 1960. He was arrested for securities fraud in 2008, pleaded guilty, and was sentenced to 150 years in federal prison in 2009. Although he claimed his business was once legitimate, he admitted that it had been nothing but a Ponzi scheme since 1991. Investigators believe Madoff's fraudulent activities extend back to perhaps the early 1980s, and it is possible that his firm was never really a legitimate business. Madoff swindled a staggering sum from his investors. The amount missing from client accounts totaled about $65 billion. The actual investment loss—money invested by clients that was never returned when the scheme collapsed—is estimated to be about $18 billion. Madoff's victims included celebrities, universities, large charities, and retirement funds.

An $18 billion swindle puts Bernie Madoff in a class by himself—at least in terms of criminals whose activities have been uncovered. So how is Bernie Madoff different from Charles Ponzi? Madoff was a lot smarter, it seems. He was able to give his fraudulent activities the cloak of legitimacy for far, far longer than Ponzi. Madoff, by his own admission, ran a Ponzi scheme for nearly 20 years; it's possible it went on for much longer. Ponzi himself rose and fell in a matter of months. Madoff's payoffs to his investors were much more modest, and thus attracted less attention, than Ponzi's outrageous returns. Signs of Madoff's fraud were there, however, and much has been made of the incompetence of government regulators through whose fingers Madoff slipped time and again over the years.

Both Madoff and Ponzi appear to have made feeble attempts to convert their fraudulent businesses to legitimate ones, at least in part to save their own skins. Converting a thriving Ponzi scheme to a legitimate business is practically impossible, though. It is the financial equivalent of trying to stop an avalanche. When something appears too good to be true, it probably is, as the old saying goes. On the other hand, as P. T. Barnum might have said, no one ever lost a dollar by overestimating the greed of the American people.

34. Let There Be Light

$$c \approx 299{,}792{,}458 \text{ m/s}$$

The Speed of Light in a Vacuum

The equation above gives the approximate value of the speed of light in a vacuum, c, in meters per second. (The value is often rounded off to 300×10^6 m/s, which is close enough for many calculations.) In U.S. customary units, the speed of light is approximately 186,282 miles per second. In other media, such as water, air, or glass, light travels at a speed less than c. The ratio of c to the speed of light in a given medium is called the refractive index, n. The refractive index for water is about 1.33, which means that the speed of light in water is about $c/1.33$, or 225×10^6 m/s.

Galileo was certainly not the first person to observe that we see the flash of a distant cannon well before we hear its boom. From observations like this, some folks concluded that whereas sound obviously travelled at a measurable rate, light must travel infinitely fast. But Galileo wasn't so sure. And so he proposed and attempted to carry out perhaps the first experiment to measure the speed of light. It may not have been very practical, but it was certainly intriguing.

Standing on hilltops a little less than a mile apart were two people holding covered lanterns. One person uncovered his lantern. As soon as the second person saw the light of the first lantern, he uncovered his lantern as well. A third person measured the time lag between the flashes of light from the first and second lanterns. Knowing that time, and the distance between the lanterns, the speed of light could, at least in theory, be calculated.

But no one is that quick! The reaction time required before the second person can uncover his lantern is far, far longer than the amount of time it takes the light to travel a mile (or even many miles). And so, given the technology of the time, Galileo's experiment wasn't especially practical, as he himself noted. Hundreds of years would pass before the development of technology that would allow a practical experiment for measuring the speed of light on the planet Earth.

Of what use is it to know just how fast light travels? For most everyday applications, the assumption that light travels infinitely fast, which was the view of Galileo's contemporaries, is perfectly acceptable. If you see a flash of lightning off in the distance, you can estimate how far away it was by first assuming that the light reached your eyes instantaneously and then counting the number of seconds until you hear the thunder that follows. (About five seconds per mile is a pretty good rule of thumb.) Knowing the actual speed of light is not necessary here. The light from a lightning flash one mile away will arrive at your eyes in about 5.4 millionths of a second. Rounding that off to zero is a pretty good approximation in this case.

But not when the distances are much greater, as they are for most astronomical observations. When you observe a planet or a star from Earth—through a telescope, say—what you see is the image of where that planet or star *used to be* when the light you are now seeing first left the planet or star. If you look at the moon, for example, what you are seeing is where the moon was about 1.2 seconds ago. In astronomical terms, the moon is not very far away; its average distance from Earth is about 240,000 miles. The sun is much farther away, on average about 93 million miles, and it takes light over eight minutes to travel that distance. The farther away a planet or star is, the longer its light takes to reach human observers on Earth. The difference between where something appears to be in the sky and where it actually is thus increases with how far away the object is. So, if you know where a planet or star is supposed to be, and you compare that with where it appears to be, you can calculate how long the light took to reach you. This is the basis for astronomical calculations of the speed of light, as first used by the Dutch astronomer Ole Rømer in 1676.

It is possible, however, to measure the speed of light right here on Earth, where the distances are obviously much shorter. A series of experimenters in the 1800s devised various mechanical means to render a version of Galileo's lantern experiment both practical and accurate. The Frenchman Jean Foucault focused a beam of light on a rapidly rotating mirror. The mirror was spinning about an axis in the plane of the mirror. Think of a flathead screwdriver, where the blade of the screwdriver represents the mirror. When you twist the screwdriver in the usual way, you are rotating its blade the same way that Foucault's mirror rotated. At one point in each revolution, Foucault's mirror was reflecting the light directly at another mirror, which was mounted in a stationary position on a distant hilltop. The light from the rotating mirror struck the stationary mirror and then bounced back towards the rotating mirror. By the time the rebounding light had completed the return trip and hit the rotating mirror, however, the rotating mirror had turned ever so slightly. The light reflecting in this manner was thus at a small angle to the original beam of light. By measuring that angle, and doing some math, you can calculate how fast the light beam is moving. Foucault's result, in 1850, was within 1% of the value accepted today for the speed of light. Not bad.

In 1880, American physicist Albert Michelson built an improved version of Foucault's experiment and was able to measure the speed of light to within a mere 30 miles per second of the value we accept today, which is close to 186,000 miles per second. In metric units, the value is 299,792,458 m/s, or very nearly 300 million meters per second.

It turns out that the speed of light is an important constant in some equations that don't really have much to do, at first glance, with how fast light is travelling. Perhaps the most famous scientific equation in history is Albert Einstein's $E = mc^2$, which relates the nuclear energy, E, that is stored in atoms to their mass, m. The constant relating E and m is c, the speed of light. It has been said that part of what got Einstein thinking about relativity, mass-energy equivalence, and other such weighty matters were the speed of light experiments of researchers like Foucault and Michelson. That would be reason enough to celebrate Galileo and all the others who searched for the true value of this fundamental constant.

35. Smarty Pants

$$IQ = \frac{\text{Mental age}}{\text{Actual age}} \times 100$$

An early equation for calculating IQ, the intelligence quotient

The subject's mental age, as measured by an intelligence test, is divided by his or her chronological or actual age. The equation predicts an IQ of 100 for a "normal" child, or one whose mental age coincides with his or her actual age.

"He's a genius with the IQ of a moron," is how the writer Gore Vidal described pop artist Andy Warhol. In what could have been a retort, Warhol famously remarked, "Don't pay any attention to what they write about you. Just measure it in inches." Genius, indeed.

In 1905, long before Andy Warhol got his 15 minutes of fame, the French psychologist Alfred Binet (1857–1911) was commissioned to develop a system for identifying children in need of special help in the classroom—the type of children we might refer to today as having learning disabilities. Binet's work was extended by others and resulted in what we now call the intelligence quotient, or IQ. In 1989, the American Academy for the Advancement of Science published a list of the 20 greatest scientific advances of the 20th century. Along with the discovery of DNA, the invention of the transistor, and heavier-than-air flight, the IQ test, the direct descendant of Binet's work, made the list.

But IQ is a two-edged sword. Intelligence measurements, the most famous of which is the IQ, are hardly without controversy, and if some organization were to solicit candidates for a list of the twenty *worst* or most destructive scientific "advances" of the twentieth century, the IQ test would be certain to garner plenty of votes. None of that is the fault of Alfred Binet.

Some of the problems with IQ can be traced back to our equation here, which wasn't put forth by Binet but by the German psychologist Wilhelm Stern, in 1912. Binet's work on intelligence measurement began around 1899 and continued until his death in 1911. The tests he developed to identify children with learning disabilities consisted of a series of tasks of increasing difficulty. Trained administrators first tested whether a child could follow a lighted match with his or her eyes or name various parts of the body. Slightly more complex tasks (repeating a three-digit number, defining simple nouns) followed. Each successful response resulted in a more complicated task. Among the most difficult tasks, for the most capable children, were repeating seven-digit numbers and thinking up words that rhymed with a given word. Through painstaking study, Binet had determined the typical abilities of "normal" children on these tests, as a function of their age. He thus knew, for example, at what rung on the ladder of increasing difficulty a typical 8-year-old child would be unable to complete the next task. This allowed him to determine a child's mental age. A 10-year-old who could complete only the tasks typical of a 6-year-old would thus be identified as a good candidate for remedial instruction.

Not long after Binet's death, Stern published our equation, in which Binet's mental age was replaced by the ratio of mental age to chronological age. A 10-year-old child who achieves a mental age of 12 thus has an IQ of 12 divided by 10 and then multiplied by 100, or 120. The difference between Stern's IQ ratio and Binet's mental age doesn't seem all that profound—but it is. Binet was careful to specify several rules for the utilization of his intelligence tests. Among these were that the tests were only for children, not adults. He also warned that the tests were not to be used for ranking normal children—just for finding those in need of remedial instruction. He cautioned as well that the tests did not measure anything inherent or unalterable about an individual child: the tests simply provided a snapshot of the child's mental development at a given point in time.

He might as well have saved himself the effort, since all of Binet's rules for the use of his tests and warnings against their misuse were repeatedly and blatantly disregarded, perhaps nowhere more so than in the United States.

Having an IQ ratio instead of just a mental age was, as noted above, part of the problem. Binet's concept of mental age (as opposed to chronological age) makes sense only for children. It doesn't make any sense to say that a highly intelligent 30-year-old has a mental age of 40, for example. But once Stern created a single unitless scale (the IQ), it was just too tempting to rank everyone, from children to adults, and from the learning-disabled to folks of normal intelligence and on up to geniuses. The U.S. Army was one of the first institutions to adopt intelligence testing for adults, with wide-ranging purposes such as admission, officer selection, and evaluation of fitness for specific duties.

Another American contribution to intelligence testing was its mass marketing. As noted earlier, Binet's tests were one-on-one, labor-intensive affairs. Not so in America, where intelligence testing became a pencil and paper test administered to large groups by a single official. With the advent of machine grading in the 1930s, intelligence testing generally became an entirely multiple-choice test.

Machine-graded testing in America is closely associated with college admissions tests such as the SAT and ACT. The original SAT, first administered in 1926, was an intelligence test. In those days, "SAT" stood for Scholastic Aptitude Test (today the initials have no official meaning). The purpose of the SAT was to try to level the playing field in a higher education system that was, especially at the most prestigious schools, dominated by members of the upper classes, who tended to gain admission regardless of their qualifications. The SAT was intended to measure scholastic aptitude, or the ability to succeed in school. The original SAT was adapted from an IQ test being used by the army and several universities. The SAT was so closely allied to IQ in those days that SAT developer Carl Brigham even published a scale for converting an SAT score into IQ. Times have changed. Today's SAT does not claim to measure aptitude or intelligence; rather, it assesses the skills that students "have developed over time and that they need to be successful in college."

The great difficulty in intelligence testing is separating out that which is innate from that which has been acquired over time. Nature versus nurture, if you will. Binet believed that intelligence varied from one child to the next but that it was not something innate or unalterable. Gradually, and after more than a few wrong turns, educators and psychologists seem to have come around to Binet's way of thinking.

36. As Old as the Hills

$$N(t) = N_0 e^{-\lambda t}$$

An Equation for Quantifying the Decay of a Radioactive Element

In this equation, $N(t)$ is the number of atoms present in a radioactive sample at some time t, and N_0 is the number of atoms present at $t = 0$. The decay constant, λ, depends on the radioactive element being considered. A commonly used parameter to describe the behavior of a radioactive material is its half-life. The half-life, which is defined as the time necessary for one-half of the atoms in a sample to decay, may be calculated as follows: $t_{1/2} = (\ln 2)/\lambda$, where $t_{1/2}$ is the half-life.

These days, global climate change grabs all of the environmental headlines. Is our planet getting warmer? To what extent is human activity responsible for climate change? These are controversial questions, and the science behind them is complex.

But back in the 1800s, there was a different question about the planet Earth on people's minds: Just how old is the planet, anyway? This was the mystery that occupied the greatest scientific minds of that era, folks like Lord Kelvin and Charles Darwin. As questions go, this one is pretty simple. There was a time, presumably, when the Earth did not exist. How long ago was that? Answering the question, as we shall see, is a little bit tougher.

A number of ancient civilizations believed that the Earth had been around forever. The advent of Christianity several thousand years ago marked a turning point of sorts in thinking about the age of the Earth. The creation story in Genesis can be interpreted to mean that the Earth may be relatively young—perhaps only a few thousand years. Back in the 1600s, the Anglican archbishop James Ussher produced a tedious study in which he worked his way backwards through the various biblical genealogies, finally arriving at 4004 BCE, the year he concluded that the Earth was created.

The age of the Earth remained mostly a religious question until about 1860. Around that time, a spirited debate began among a veritable who's who of top scientists from a wide range of disciplines. Physicists, geologists, astronomers, biologists, and others all weighed in on the subject, and each branch of science had its own method or methods for measuring and calculating the age of the Earth.

To measure the age of the Earth, you need a clock. That is, you need something about the Earth that varies regularly with time. Lots of different clocks have been proposed to measure the age of the Earth. Various astronomical clocks depend on the motion of the Earth, the moon, and other planets. There are geological clocks, too; these might depend on the rates at which the layers in the Earth's crust build up over time. For a while, there was much interest in a salt clock, which depended on the rate at which salt was building up in the oceans of the world.

And then there was the thermodynamic clock. One of the most important players in the age of the Earth debate in the 1800s was the venerable Lord Kelvin, William Thomson. Kelvin, one of the founders of the modern science of thermodynamics, approached the problem—no surprise—thermodynamically. It was well known back then that the Earth gets warmer as you dig down from the surface. On average, for every 50 feet you descend, the Earth gets warmer by about 1°F. So the Earth is just a giant body cooling off very, very slowly. A medical examiner, as every fan of crime fiction knows, can sometimes estimate the time of death of a murder victim by measuring body temperature. When a person dies, the body begins to cool off at a rate that is well documented. Provided the body is examined quickly enough, before it has cooled to ambient temperature, time of death can be estimated with reasonable accuracy using this method.

Kelvin's approach was similar. In the beginning, the Earth was a mass of molten rock, and it's been cooling ever since. He set about to estimate the rate at which the planet would cool, taking into account all the energy flows into it (such as sunlight) and out of it, as it cooled. His estimate placed the age of the Earth at somewhere between 20 million and 400 million years. Kelvin's acknowledged that he'd had to make quite a few assumptions in his calculations; thus, the wide gap between his upper and lower bounds on the age of the Earth. One important energy source not taken into account in Kelvin's equations was the massive amount of energy constantly being given off by all the radioactive elements in the Earth. This tremendous internal heat source in our planet effectively slows its cooling rate, and this, among other things, caused Kelvin to conclude that our planet is much younger than it really is.

Unfortunately for Kelvin, the discovery of radioactivity by Becquerel and Roentgen in the 1890s postdated Kelvin's work on the age of the Earth. Later researchers—Marie and Pierre Curie and Ernest Rutherford prominent among them—formulated the theories that govern the behavior of radioactive elements. These theories would eventually provide the key to unlock the mystery of the age of the Earth.

Radioactive elements are unstable. They decay spontaneously into other elements, all the while emitting energy in the form of radiation. The rate at which radioactive elements decay into other elements is exceedingly precise. If you can measure this decay rate, you have at your disposal an exceptionally accurate clock—and it is this clock that was eventually determined to be the most reliable and precise method for determining the age of the Earth.

In our equation above, $N(t)$ is the number of atoms present in a radioactive sample at some time t, and N_0 is the number of atoms originally present, at time $t = 0$. The "decay constant," λ, depends on which type of element is under consideration; all radioactive elements decay at different rates and thus have different values of λ. Rutherford began to figure this all out in 1905, and others have improved on his work.

Radioactive dating methods have been applied now for about a hundred years. Rocks exceeding 3.5 billion years of age have been found on all the Earth's continents. Meteorites with ages of about 4.5 billion years have been found on Earth. Moon rocks between 3.5 and 4 billion years are common. The best science available today thus places the age of the Earth at 4.5 billion years, plus or minus 1%. Mother Earth looks pretty good for her age, all things considered.

37. Can You Hear Me Now?

$$RT_{60} = \frac{0.049V}{S_e}$$

Wallace Sabine's Equation for Reverberation Time

RT_{60} is the time necessary for a sound to decay by 60 decibels (at which point the sound is essentially inaudible). V is the volume of the room (in cubic feet) in which the sound is made, and S_e is the effective sound absorbing area of all the surfaces in the room. Values of S_e, which vary with the frequency of the sound, have been determined and tabulated for all sorts of materials. For a frequency of 1000 Hz, the S_e of a wooden floor, for example, equals 0.07, whereas a heavy carpet backed by foam rubber measures 0.69.

It was a problem nobody wanted to work on. In 1895, Harvard University opened the Fogg Art Museum, which included the Fogg Lecture Hall. But the new hall quickly gained a reputation as a place to avoid because its acoustics were so poor that it was difficult for anyone in the audience to understand what the lecturer was saying. Having spent a lot of money on this facility, Harvard was anxious to fix the problem. And so the university turned to its renowned physics department. One by one, however, the distinguished faculty members of Harvard's department of physics declined to help. Having heard the acoustics in Fogg Hall for themselves, perhaps they figured it was a lost cause.

The problem of the acoustics of Fogg Lecture Hall eventually landed in the capable hands of one of the physics department's youngest professors, Wallace Clement Sabine (1868–1919). Sabine is now widely considered the father of architectural acoustics. In 1895, he was just a young lecturer at Harvard who had no particular experience with acoustics.

What makes one lecture hall sound better than another? How do the acoustics of a music hall differ from those of a lecture hall? Is it possible to design a hall so that it sounds good for various types of instrumental and vocal music as well as for the spoken word? Before Wallace Sabine took on the Fogg Lecture Hall, there were no quantitative answers to these questions. Now there are.

Sabine brought a quantitative, scientific approach to architectural acoustics. One of his fundamental contributions was the development of the concept known as reverberation time. Stand in the middle of a room, and clap your hands together. How long did it take for the clapping sound to die out? In the furnished living room of a typical home, the reverberation time might be somewhere around one second. When you clap your hands, the sound waves you create travel outwards, bouncing off the walls, floor, ceiling, and other objects. Some of those reflected waves make it back to your ears. As time passes, all of those sound waves bouncing around start to die out, and eventually you can't hear the original sound you created.

Sabine was able to quantify the relationship between the size of a room, the materials from which it was constructed, and the reverberation time. Imagine repeating your handclap experiment in the same room in your house, except that now the room is unfurnished—no carpeting, curtains, chairs, tables, or couches. The reverberation time will now be much longer—perhaps several seconds longer than in the same room when it was furnished. Realtors prefer to show furnished homes to potential buyers. Unfurnished rooms appear cold and unwelcoming. But the harsh acoustics of unfurnished rooms are an important turnoff to potential buyers, too. Furnished rooms are more inviting to the eye and to the ear.

Fogg Lecture Hall, in 1895, wasn't particularly inviting to the ear. In early experiments there, Sabine measured reverberation times of over five seconds! Five seconds is longer than it would take the average person to read this sentence aloud. If the person reading that sentence aloud were doing so in a room with a reverberation time of five seconds, it's easy to see how difficult it would be to understand him or her, since the reverberations from earlier words would be constantly interfering with later words.

A reverberation time of five seconds or more isn't always a bad thing. The magnificent cathedral of Notre Dame de Paris has a reverberation time as long as 8.5 seconds. This gives the cathedral's huge pipe organ a full, rich sound that it wouldn't have in an acoustically quieter environment. Just don't try to give a speech there.

So that was the problem at the Fogg Lecture Hall. Working at night, when the lecture halls were empty, Sabine ran a series of experiments in Fogg, with its poor acoustics, and in other lecture halls at Harvard, including some renowned for their excellent acoustics. Employing a portable pipe organ, he would play a note, and then record how long it took for the sound to die out. His only instruments were a stopwatch and his ears.

Sabine had help, though. A small army of students helped him reconfigure the furnishings in Fogg Hall and the other rooms he investigated. He made measurements with and without rugs on the floor, with and without cushions on the chairs in the rooms, and with varying numbers of people seated in the chairs. Among his findings was this: the presence of a single person seated in the hall lowered the reverberation time as much as six seat cushions.

Eventually, Sabine was able to quantify the relationship between reverberation time, the size of a hall, and the amount of sound-absorbing material it contained. This is our equation above. RT_{60} is the reverberation time necessary for a sound to decay by 60 decibels, or essentially to become inaudible; V is the volume of the room in cubic feet; and S_e is the effective sound absorbing area of all the surfaces in the room. Determining S_e for a given room takes some doing. First of all, different materials absorb sound differently. Put a cushion on a wooden chair, and the chair will absorb more sound. Put a person on that chair, and now it will absorb even more sound. Sabine was able to quantify and tabulate the "absorption coefficients" of all kinds of materials. Knowing those, and having measured the surface areas of the various materials in a room, he could calculate S_e.

Applying this new knowledge to Fogg Lecture Hall, Sabine made recommendations for installation of sound-absorbing materials that greatly reduced the reverberation time in the hall and made it acceptable for its primary intended use—lecturing. Sabine also quantified what he believed were ideal reverberation times for different purposes. Longer reverberation times, perhaps two to three seconds, are better for instrumental music performances—this reverb time creates a richer, fuller sound. Vocal music, such as opera, requires somewhat shorter reverberation times. The spoken word, as in a lecture hall, is best in a hall with an even shorter reverberation time of about one second.

Performance halls are most effective, however, when they are multipurpose. On Monday night, such a hall might feature a lecture by an author or scholar, on Tuesday night a play, on Wednesday a string quartet, and on Thursday through Sunday an opera. For this reason, most performance halls can be reconfigured to change their reverberation time. One way this is commonly done is with retractable curtains on the side and back walls of the hall. Such curtains might be fully deployed for a spoken word performance, fully retracted for instrumental music, and left somewhere in between for an opera.

After his success with Fogg Lecture Hall, Sabine was hired as an acoustical consultant on the design of Boston's Symphony Hall, which opened in 1900. It was the first concert hall to be designed using quantitative acoustics, and more a century later, it retains a worldwide reputation as a splendid acoustical environment.

38. Decay Heat

$$\frac{P(t)}{P_0} = At^{-a}$$

Heat Generated after Shutdown by a Nuclear Reactor as a Function of Time

The equation above represents one way of calculating the rate at which heat is being generated by a nuclear reactor, as a function of time, after the reactor has been shut down. $P(t)$ is the rate at which heat is being produced by the reactor at some time t after shutdown. P_0 is the maximum power that reactor can produce—the maximum rate at which it can produce heat. The ratio of $P(t)$ to P_0 decays exponentially as time t increases, as shown in the equation. The decay rate depends on two constants, A and a.

When the March 11, 2011, earthquake and tsunami struck northeastern Japan, the reactors at the Fukushima nuclear power plant did exactly what they were designed to do: they shut themselves down. Unfortunately, that was only the beginning of their problems.

Nuclear power plants are, in several important ways, not that different from conventional electrical generation power plants, such as those powered by burning coal or natural gas. In each case—whether the plant is powered by nuclear reactions, coal, or natural gas—tremendous quantities of heat are generated and then used to boil massive amounts of water. The steam thus created spins a turbine connected to an electrical generator. What differs among these plants is merely how all of that heat is created.

Burning coal or natural gas to create heat is pretty simple. Combustion requires a fuel, a source of oxygen, and a spark to ignite it. A fire started in this way will keep burning as long as the fuel and the oxygen hold out. To shut down a combustion reaction, you simply cut off the fuel. Think of boiling a pot of water on a natural gas–powered kitchen range. You fire up the gas burner, and a few minutes later, the water in the pot begins to boil. When you turn off the gas, killing the flame, the water stops boiling almost immediately.

Come back an hour or so later, and the water in the pot will have cooled off to room temperature.

Now imagine that instead of a natural gas burner, your kitchen range was powered by a small stovetop nuclear reactor. In a nuclear reactor, the idea is to bring together enough nuclear fuel—things like uranium 235—to cause a controlled nuclear reaction.* In a controlled nuclear reaction, large atoms, such as uranium 235, split apart to create several smaller atoms. Each time this happens, several neutrons, which are particles found in the nucleus of an atom, are also liberated. These neutrons blast into other uranium atoms, causing them to split apart, and the whole reaction becomes self-sustaining. Such reactions give off prodigious amounts of heat, which you could use to boil the water on your stovetop.

To control all of these nuclear reactions, you have to control what happens to all the neutrons that are liberated. Nuclear reactors thus contain control rods, which are made of materials that absorb neutrons and keep them from participating in other nuclear reactions. Moving the control rods in or out of the nuclear reactor is a little bit like turning the gas down or up on a gas burner. Move the rods out, and you get more nuclear reactions and more heat. Move the rods in, and you get fewer reactions and less heat.

So, now you've got the water boiling in the pot on top of your little stovetop nuclear reactor. Inserting the control rods all the way shuts down the reactor; doing so essentially prevents the liberation of more neutrons to sustain the nuclear reaction. But the amount of heat being created inside your reactor, unlike that created by the gas burner, does not drop to zero. A large gas burner on your stove might create heat at the rate of 10,000 Btu per hour. This converts to 2,928 watts—let's call it 3,000 watts. When you turn off a gas burner, it goes, in a fraction of a second, from creating heat at the rate of 3,000 watts to creating zero watts.

Let's say your stovetop nuclear reactor is also a 3,000-watt model. When you turn it off, it goes from 3,000 watts of heat to about 210 watts one second later. Four hours later, it would be producing about 30 watts of power, or 1% of its maximum heat-generating power. A month after shutdown, the reactor would still be producing maybe 15 watts, or 0.5% of its max power. And there is absolutely nothing you can do to stop this heat from being created.

*Starting up a nuclear reactor involves more than just bringing together enough nuclear fuel. "Startup neutron sources" are typically also required in order to jumpstart the whole process.

This post-shutdown heat is called decay heat, and our equation shows one way it can be calculated. The left-hand side of the equation, $P(t)/P_0$, is the ratio of the power produced by a shut-down reactor after a certain amount of time, t, to its maximum power, P_0. On the right-hand side, A and a are constants, and t is the time. The power ratio thus drops off very quickly at first, but then remains at a small but not zero value for a very long time.

A power ratio of half of 1% after one month doesn't seem like very much, and it isn't—as long as we're talking about a stovetop burner. But if you multiply the heat generated by our stovetop burner by, say, about 300,000 times, you get the amount of heat that might be generated by a nuclear reactor. *Now* when you take 1% or 0.5% of that huge number, you're talking about a *lot* of heat.

Four hours after they were shut down, each of the Fukushima reactors was still producing heat at the rate of something like 30 megawatts (30 million watts). A week after shutdown, heat was still being produced at about 10 megawatts in each reactor. Safely getting rid of that heat, in the face of all manner of natural and manmade calamities, is crucial. It is perhaps the most challenging aspect of designing a nuclear power plant. The tsunami waves in Japan overwhelmed the Fukushima power plants, destroying the reactors' cooling systems. The heat still being produced by the reactor was not being removed, and the reactors overheated, resulting in explosions, releases of radioactive material, and so forth.

Why can't you simply turn a nuclear reactor all the way off, the way you do a gas burner? It's because even though all the neutron-generating reactions have been effectively stopped when you shut down the reactor, the elements (uranium and others) in the reactor continue to give off heat through a separate process called radioactive decay. This is a completely natural phenomenon. It's why, for example, some folks' houses might contain dangerous levels of radioactive radon gas, which naturally filters up through the earth and into houses in some parts of the world. Because there is so much radioactive material gathered together inside a nuclear reactor, the natural and unstoppable process of radioactive decay produces significant amounts of heat in a relatively small space. When a reactor is shut down, *that heat must at all costs be continuously removed*, lest the reactor subsequently melt down or explode.

Nuclear physicists and nuclear engineers are trained to think in terms of probabilities—the probability that a certain nuclear reaction will happen, the probability that the neutron thus liberated will cause another reaction, and so on. Well, the probability that a nuclear reactor will stop generating heat when it is shut down is *zero*. But what about the probability of the total failure of all the various redundant cooling systems designed to get rid of all of that decay heat? That probability can be very small, but it can never be zero. Fukushima provides an example of this. What is needed is a reactor design that is safe even when all of its cooling systems have failed. Nuclear engineers tell us that this is possible and that such designs exist. The cost of building new reactors, of any kind, however, is daunting. And in the wake of Fukushima, humankind's appetite for nuclear power seems to have greatly diminished.

39. Zero, One, Infinity

$$N = R^* \times f_p \times n_e \times f_\ell \times f_i \times f_c \times L$$

The Drake Equation

The Drake equation is used to estimate N, the number of civilizations in the Milky Way with which we might be able to communicate.

R^* = the average rate of star formation per year in our galaxy
f_p = the fraction of those stars that have planets
n_e = the average number of planets that can potentially support life per star that has planets
f_ℓ = the fraction of the planets that actually develop life at some point
f_i = the fraction of the life-creating planets that develop intelligent life
f_c = the fraction of civilizations that develop a technology that emits detectable signs of their existence into space
L = the length of time over which such civilizations emit detectable signals into space

To say that the universe is a big place is to state the obvious. The universe is so big that when you are counting things, instead of going "zero, one, two, three, . . ." you can pretty much go "zero, one, infinity." That's because in the universe, once you find a second example of something, you can pretty much be sure that the universe contains a really, really big number of whatever item it is that you are counting. Not an infinite number, to be sure, but a really big number nonetheless.

Consider, for example, planets. We live on one—the Earth. A long time ago, stargazers discovered, one by one, the other planets in our solar system. The concept of "zero, one, infinity" says that, once that second planet was discovered, you could safely conclude that there are an incredibly large number of planets in the universe. That's how big the universe is.

And so the universe contains so many planets that you could scarcely count them all. But what about intelligent life? Among all those planets, how many have developed intelligent life? We know of only one such planet—the one we live on. But if we haven't been able to count to two yet, it's not for lack of trying. And once we get to two—if we ever do—we'll know that there are a huge number of planets with intelligent life.

Not that many years ago, searching for intelligent life in outer space was only for kooks. Now, it has become an accepted subspecialty within astrophysics. One of the early contributions in this area is our equation here. Called the Drake equation, it attempts to predict how many examples of life on other planets there might be in our galaxy, the Milky Way, with which we might be able to communicate.

The Drake equation has lots of variables, as described above. The goal of the equation is to estimate N, the number of civilizations in the Milky Way with which we might be able to communicate. Frank Drake is an astronomer and astrophysicist. He published his equation at a Search for Extraterrestrial Intelligence (SETI) conference in 1960. Originally, Drake saw the equation not so much as a means to actually calculate the number of civilizations out there, but more as a means of organizing and understanding our ignorance concerning the likelihood of extraterrestrials.

With encouragement from his colleagues, Drake made his first estimate with the equation in 1961. Using an assortment of values—essentially guesses based on very little data—Drake calculated that $N = 10$. Thus, he predicted that there are 10 civilizations within our galaxy with which communication might be possible. Many others have made predictions with Drake's equation. In 1966, Carl Sagan, a huge proponent of the equation, used much higher estimates of the values of the equation's variables to calculate the number of communicating civilizations in the Milky Way to be around one million. Sagan promoted the equation so strongly that it is sometimes mislabeled as the Sagan equation. Others, less optimistic than Sagan, have used the equation to estimate N to be much less than one.

Our theme of "zero, one, infinity" comes through in these predictions—especially when you recall that Drake's equation takes into account only a single galaxy, the Milky Way. Conservative estimates of the number of galaxies in the universe put the figure at 125 billion.

The Drake equation helps bring into focus several issues related to the search for extraterrestrial life. The Fermi paradox, for example, points out the apparent contradiction between the high probability of intelligent life within our galaxy (as predicted by some using the Drake equation or other means) and the utter lack of positive results from attempts to communicate with that life.

The lack of extraterrestrial communication is traditionally known as the "Great Silence" or "Silencium Universi." The Drake equation sheds some light on this through the variable L, the length of time over which intelligent civilizations release detectable signals into space. Humans have been sending detectable signals (such as radio waves) from Earth into space for not much more than a century. This, on a planet that is over 4.5 billion years old! It is entirely possible that civilizations on other planets were born, lived, and died out billions of years before humans evolved on Earth.

In addition to the time problem, there is also a space problem. If you assume that no one has figured out how to send signals faster than the speed of light, then it is possible that other intelligent civilizations are just too far away to communicate with us, especially when you consider that we haven't been around very long.

Scientists who study all of this tend to fall into two camps, which we might label "zero" and "infinity." The "infinity" folks believe that because the universe is so big and has been around so long, a really huge number of intelligent civilizations have arisen. Because of that, they further believe that it is only a matter of time before we manage to communicate with some of them. The "zero" folks counter that the conditions that gave rise to intelligent life on Earth are exceedingly rare—even in a universe with such a huge number of planets. When the constraints of time and space are factored in, they believe it is likely that we will never communicate with extraterrestrial intelligent life.

In the search for intelligent civilizations, we are stuck at one—our own. When will we get to two? Or will we ever get there at all?

40. Terminal Velocity

$$F_{\text{aero}} = 0.5\,\rho C_d A v^2$$

The Aerodynamic Drag Force on a Body Moving through Still Air

This equation represents the aerodynamic drag force, F_{aero}, on a body moving with velocity v through still air (a wind velocity of zero). The density of the air is ρ, the frontal area of the body is A, and the aerodynamic drag coefficient of the body is C_d. The drag force increases with the square of the velocity. The quantity $C_d A$ is called the effective frontal area of an aerodynamic body, which takes into account both the body's size (A) and its aerodynamic "slipperiness," C_d.

Those of us who've never jumped out of a perfectly good airplane often wonder why others do. It must be for the thrills, among those being the sensation of speed. Jumping from altitudes typically between 12,000 and 14,000 feet, a falling skydiver approaches the earth at a maximum speed of about 125 miles per hour (close to 200 kilometers per hour) when falling in the classic spread-eagle position—parallel to the earth with arms and legs spread wide. This is the so-called "terminal velocity," the maximum speed a body (animate or inanimate) falling through the atmosphere can achieve. When falling in an upright position (more like a bullet), the terminal velocity of a skydiver increases to about 200 miles per hour. Our equation explains why, as we shall see presently.

The force accelerating a body downward is its weight. The force resisting that downward motion is known as the aerodynamic drag force, or F_{aero} in our equation here. The faster a body falls, the greater the drag force becomes. When the drag force finally equals the weight of the body, the body stops accelerating, having reached its terminal velocity.

Terminal velocity varies with lots of things, one of them being the configuration of the body (spread eagle versus bullet, as discussed above). And while falling through the air at 125 miles per hour may seem fast, it's nothing compared to what Felix Baumgartner accomplished on October 14, 2012, in the Red Bull Stratos Project.

Red Bull is best known as a company that makes energy drinks, but that's not all it does. The Red Bull name is also linked to the company's connection with extreme sports. Red Bull sponsors a lot of athletes and events, but there's never been anything quite like the Stratos Project. "Stratos" is short for "stratosphere," and the idea was simple: have a skydiver jump from the stratosphere and live to tell about it. Felix Baumgartner jumped from approximately 128,100 feet (about 24.3 miles, or 39 kilometers) above the Earth's surface near Roswell, New Mexico. In the process, he appears to have set three world records. One record was for ascending in a helium balloon (highest manned balloon). Another record fell about 42 seconds after he jumped, when he become the first human to break the sound barrier during free fall. Mach 1 at that altitude is around 690 miles per hour, or 1,100 kilometers per hour. While the sound barrier is frequently broken by airplanes, it had never been achieved only under the force of gravity. After breaking the speed of sound, Baumgartner's top speed eventually reached about 834 miles per hour (1,343 kilometers per hour). His third record was for the highest-altitude jump.

Falling from the stratosphere is fraught with peril. The stratosphere is the second layer of the Earth's atmosphere. The layer closest to Earth, the troposphere, reaches up to an altitude of about 6 miles (10 kilometers). The stratosphere extends from there on up to an altitude of about 31 miles (50 kilometers). The stratosphere is physically different that the troposphere because the air temperature begins to increase with height, whereas a height increase in the troposphere leads to cooler air. The stratification of warmer air at higher altitudes and cooler air at lower altitudes provides the stratosphere with the root of its name. Among lots of other things, Baumgartner had to contend with these temperature gradients. At the beginning of his jump, the temperature was around −10°F (−23°C), but as he descended through the stratosphere, the temperature fell to about −40°F (−40°C).

The atmospheric pressure and air density at 100,000 feet are not to be trifled with, either. Air at this altitude contains only 1% of the oxygen it contains at sea level—not nearly enough to survive without special breathing equipment. It is because the air is so thin, however, that Baumgartner had the chance to reach supersonic, record-setting speeds. In the troposphere, the drag force is far too great for a human to reach such high speeds owing to gravity alone.

In our equation, F_{aero} is the drag force, C_d is the drag coefficient, ρ is the air density, v is the velocity, and A is the cross-sectional area that the falling object presents to the wind—the so-called frontal area. The drag force thus decreases proportionally with the air density. (This is one reason jet airliners cruise at such high altitudes, where the reduction in air drag dramatically increases their fuel efficiency.) If the air at stratospheric heights is only 1% as dense as the air at sea level, the drag force for a body falling through that air is only 1% as great as it would be at sea level. Terminal velocity in such thin air is thus much greater than it would be in air of normal density.* The low air density in the stratosphere was both a necessity, in order to break the speed record, and an extreme danger.

The low air pressure in the stratosphere can also cause ebullism, a painful, potentially lethal condition in which the fluids inside the body turn to gas. The fluids in your body can literally boil at such low pressures. David Clark was in charge of manufacturing the full pressure suit and helmet necessary to protect Baumgartner. Clark's company designed the equipment used by Joseph Kittinger (Baumgartner's mentor) in his famous 102,800-foot (31,330-meter) jump back in 1960. Kittinger held the record for the highest jump until Baumgartner came along, over 50 years later. Kittinger reached free fall speeds of over 600 miles per hour, coming close to but not breaking the sound barrier. The suit designed for Baumgartner by the David Clark Company made history as the first space suit created for a mission not sponsored by a government.

The Stratos Project was not merely an expensive and dangerous publicity stunt. It contributed knowledge to the scientific community about the human body and its ability to function at supersonic speeds. Control of the body is crucial at such high speeds in order to avoid violent spins and vibrations. Very early in the jump, Baumgartner began to spin uncontrollably, but he was somehow able to recover. Otherwise, he would have been forced to deploy a drogue chute, dashing any chance at a speed record. Someday, astronauts may be able to eject from their high-altitude missions safely using the technology and strategies employed during the Stratos Project. The rest of us will have to content ourselves with the relatively mundane thrills associated with conventional skydiving.

*Our equation also shows us why a skydiver in a vertical position achieves a higher terminal velocity than one in the more classic spread-eagle position. The frontal area, A, is much larger when the diver is spread-eagled, and so is the drag coefficient C_d. These increase the drag force and thus lower the terminal velocity.

41. Water, Water, Everywhere

$$F_{mag} = V_{sphere} M_{sat} \nabla B$$

Force Exerted on a Particle by a Magnetic Field

When a small particle of ferromagnetic material (a material strongly attracted to a magnet) is exposed to a magnetic field, a force, F_{mag}, is exerted on that particle. F_{mag} may be calculated from the volume, V, of the particle (here assumed to be spherical) and the magnetic saturation, M_{sat}, of the material when the particle is exposed to a magnetic field of gradient ∇B.

There are those who believe that the next world war, when and if it occurs, will be fought over access to drinking water. More than seven billion humans share our planet now, and there is, on average, plenty of safe, fresh water to go around. Unfortunately, all of that water isn't distributed "on average." The billions of gallons of water, for example, that are used annually to irrigate all the golf courses in the deserts of southern California could save many lives around the world, but getting that water to the folks that need it would require far more than good intentions.

Water problems range from simply not having enough of the stuff to having plenty of water but with plenty of problems. Drinking water can be contaminated by many things, including pesticides, fecal matter, bacteria, industrial pollution, and minerals that occur naturally in the ground. Problems with fresh water are as different as the countries and regions in which they are found. Here, we focus on just one country and just one water problem: Bangladesh and arsenic.

Not counting city-states like Hong Kong and Singapore, Bangladesh is the world's most densely populated nation. About 142 million Bangladeshis are crammed into a relatively small country about the size of the state of Iowa (population 3 million). There are nearly 2,500 persons per square mile in Bangladesh. The United States, in contrast, is a huge, empty country, with only 83 inhabitants per square mile. The world's most populous country, China, has 363 persons per square mile, while in India the figure is 952.

Historically, Bangladeshis relied on surface water instead of well water. To combat widespread problems with bacterial contamination of surface water that grew worse towards the end of the 20th century, Bangladesh began a concerted effort to drill water wells. Unfortunately, it just traded one problem for another.

Arsenic—an odorless, tasteless, poisonous element that occurs naturally in the Earth's crust—contaminates much of the well water of Bangladesh and is estimated to be killing something like 3,000 Bangladeshis each year. An estimated two million Bangladeshis currently have toxic levels of arsenic in their bodies. Of the 64 districts in Bangladesh, 59 have well water with arsenic levels above limits set by the U.S. EPA (Environmental Protection Agency).

In high concentrations, arsenic is deadly; it was once widely used as a rat poison. In the lower doses typical of contaminated drinking water, arsenic's effects are slower and more insidious. Drinking arsenic-contaminated water can lead to many types of cancer, as well as heart disease, skin lesions, and various kinds of infections.

Arsenic is the 52nd most common element in the Earth's crust. Like most elements, however, its distribution is hardly uniform; and Bangladesh is far from the only place with serious arsenic problems in its water. There are, for example, plenty of places in the western United States with lots of arsenic naturally occurring in the soil. Why don't folks in those places suffer from problems related to arsenic-contaminated drinking water? They certainly can, but they generally don't because the problem is typically limited to well water, and most people in those regions don't drink well water. Those who do are advised to have the water checked out carefully to make sure the arsenic levels are below EPA limits. If arsenic is found in a well that is used for drinking, the arsenic can be safely removed. Thus, in a prosperous country like the United States, arsenic problems in drinking water are relatively rare.

Things are different in Bangladesh. When you combine widespread arsenic contamination with the world's most densely populated nation and extreme poverty, the result is something of a perfect storm of water quality issues. Accurately measuring arsenic concentrations in well water in Bangladesh and then removing the stuff is a lot more difficult than you might imagine. Accurate measurement of arsenic in the laboratory involves expensive, bulky instruments that aren't especially amenable to field-testing. Bangladesh is estimated to have something like 11 million water wells. Any sort of practical testing regimen has to be inexpensive, quick, portable, and reasonably accurate. Once dangerous levels of arsenic in water have been identified, the arsenic must be removed before the water can be safely consumed.

One effective technique for removing arsenic from water involves first reacting the arsenic with a chemical to create water-insoluble compounds that can then be removed by filtration. This works well, but it results in relatively large quantities of waste materials that must be safely disposed of.

A new technique for arsenic removal that could reduce the amount of waste involves the use of magnetic nanocrystals. This technique has its roots in processes sometimes used in the mining industry. Researchers at Rice University found that arsenic in water will bind with tiny crystals of iron oxide that are added to contaminated water; these nanocrystals (along with the arsenic) can then be removed from the water using magnetic separation techniques. Because the process requires relatively small amounts of the iron oxide (0.5 grams per liter of water), the volume of waste material to be disposed of is much less than for traditional reaction and filtration techniques.

Our equation here refers to the magnetic separation technique itself. The water containing the iron oxide nanoparticles is passed through something called a high-gradient magnetic separator—essentially a metal tube packed with steel wool and subjected to a magnetic field. The magnetic field causes the iron oxide particles to be attracted to the tube and the steel wool, effectively removing them from the water.

The force that is attracting the iron oxide particles to the column and the steel wool is quite complex. One equation that can be used to understand this phenomenon models this force as a function of the volume of the particle (which is assumed to be spherical); its magnetic saturation, M_{sat}; and the gradient of the magnetic field, ∇B. Researchers found that the optimum size for the nanocrystals was about 12 nanometers. (The thickness of a human hair is about 2,000 times as large as this.)

Will this technique prove practical for the removal of arsenic from well water in places like Bangladesh? Time will tell. In the meantime, turn on the tap, fill a glass with water, and take a drink. If you live somewhere where you can safely do so without thinking twice, you're luckier than you might imagine.

42. Dog Days

$$\text{Dog years} = 7 \times \text{age}$$

Comparison of a Dog's Age to That of a Human

The age of a dog, when multiplied by 7, gives a very rough approximation, in dog years, to the corresponding age of a human being.

In the Darwinian struggle for survival, why do some species age so much more quickly than others? While human beings are thought to be the longest-living land mammals, plenty of aquatic species, including some mammals, live much, much longer than we do. Some whales, for example, have lived more than 200 years. Back on land, a giant tortoise in an Indian zoo died in 2006 at the age of 255 years. But at the other end of the spectrum are all manner of creatures whose time on Earth, on average, is not very long at all.

Consider man's best friend, the dog. Some breeds of dogs live less than 10 years, and even the longest-lived breeds don't last, on average, much more than 14 years. A 20-year-old dog is probably rarer than a 100-year-old human.

If you own a dog, and even if you don't, you are probably familiar with the concept of dog years. Our equation here illustrates this simple concept. Because dogs age so much more quickly than people, you can make a rough comparison between the ages of dogs and people by multiplying the dog's age by 7 to get "dog years." For example, a 2-year-old dog has 2×7, or 14, dog years—a teenager in human terms. An 11-year-old dog is 77 in dog years—pretty darn old.

One obvious problem with this is that different breeds of dogs have wildly different life expectancies. According to pets.ca, Irish wolfhounds live on average only a little over 6 years, while miniature poodles live more than twice as long—almost 15 years on average. Multiplying by 7 to get dog years is thus, at best, a very rough approximation.

Dogs have shorter life spans and age more quickly than human for reasons that aren't entirely understood. The science of aging is complex, although our understanding of it is getting better all the time. Even the term "aging" is not as simple as it might at first appear. Simply defined, aging is "the process of growing old." But then what is "old"? Peter Medawar, who won the 1960

Nobel Prize in physiology or medicine, defined aging as "the collection of changes that render [living creatures] progressively more likely to die." And indeed, if you look at a graph of human mortality rates (number of deaths per 1,000 people) versus age, you will see a steady increase in mortality beginning at around the age of sexual maturity.

So why are humans, or dogs for that matter, "progressively more likely to die" as they grow older? Just what kinds of things constitute the "collection of changes" that takes place?

First of all, there are intrinsic and extrinsic factors in aging. Our hair, for example, turns gray and then white as we get older, mostly because of intrinsic aging. The pigments that give our hair its color are, for complex reasons, gradually left out of the hair shafts as they grow. The hair of a typical Caucasian, by age 50, is 50% gray. The pigments that previously made it black or brown or red have largely disappeared. By the time that hair has turned white, those pigments have completely vanished.

The aging of skin, in contrast, is a combination of intrinsic and extrinsic factors. After age 20, the skin your body produces contains less and less collagen each year—this is intrinsic aging. Collagen is a protein found in all sorts of cells in the body. It is especially prevalent in skin, where it provides firmness, texture, and resilience. As the skin's collagen content naturally diminishes over the years, so do the skin's youthful qualities. But the aging of skin is also crucially dependent on extrinsic factors. Sunlight is the most important of these, but tobacco use and other factors matter as well. The ultraviolet radiation in sunlight damages skin cells in complex ways that range from largely benign and temporary (suntan) to much more serious and permanent (wrinkling, thinning, skin cancer).

Aging has intrinsic and extrinsic causes, and so does mortality. When a zebra gets eaten by a lion, that's extrinsic mortality. Intrinsic mortality, then, is the death rate that a species would exhibit under environmental conditions that are more or less benign—as in a zebra that is well cared for in a zoo. Biologists have long suspected that there is a relationship between intrinsic and extrinsic mortality. For a long time, it was thought that a high rate of extrinsic mortality was probably tied to a high rate of intrinsic mortality. Why should a species be genetically programmed to live a long time (low intrinsic mortality) if it was probably just going to get eaten while it was still very young? More recent research has shown that the relationship between intrinsic and extrinsic mortality is not quite that simple. In order to survive, a species has to be able to reproduce, avoid predators, and survive other extrinsic factors, such as weather extremes. How all of this relates to the genetically programmed ability of a species to live for a long time—to its intrinsic mortality—is a complex question indeed.

The science of biological aging is known as senescence. Some species are negligibly senescent: they do not age the way that almost all other living beings do. Negligibly senescent creatures—including certain species of rockfish, sturgeons, and tortoises—do not exhibit reductions in reproductive ability or other biological functions as they get older, and their death rates do not increase with age. Are such creatures "biologically immortal"? That is, do they die only from extrinsic causes? Such a conclusion is controversial, but nonetheless, there remains a great deal of interest in why such species live so long and exhibit so few signs of aging. A better understanding of all of this could, someday, have implications for human medicine.

For the time being, however, we must confront our own mortality and make the best of the all too brief time we are allotted on this Earth. Which may include sharing a home with a canine companion. Believe it or not, some academic studies show that dog owners actually live longer than the rest of us—even if, sadly enough, their dogs do not.

43. Body Heat

$$j^* = \sigma T^4$$

The Stefan-Boltzman Law

The total energy per unit time (or power) and per unit area emitted by a so-called "black body" (an ideal emitter of radiation) across all frequencies is proportional to the absolute temperature of the body raised to the fourth power. In SI units, the energy per unit time per unit area, j^*, is measured in watts per square meter and the absolute temperature, T, in Kelvins. The Stefan-Boltzman constant, σ, is approximately 5.67×10^{-8} W/(m^{2}K^4).

They don't call it the "miracle of life" for nothing. Many different body systems have to work right to keep a human being alive. Not the least of them is the body's ability to regulate its own temperature. We are optimized for a core temperature of 98.6°F (37°C), and prolonged temperature changes of more than a few degrees in either direction can have disastrous consequences. Energy flows into and out of the human body in a variety of ways, and all of them have to be accounted for if we are to maintain our equilibrium temperature. One of the most important means by which energy leaves the human body is infrared (IR) radiation, sometimes referred to as heat waves, and that is our focus here.

Above the temperature of absolute zero (−273°C or −460°F), everything emits radiation. This includes not only the sun, with a surface temperature of about 5500°C (9932°F), but also the cup of coffee on your kitchen table and even the apple sitting next to your coffee. That the sun emits radiation owing to its fearsome temperature is not a surprise—we can both see that radiation and feel it. Minding its own business at room temperature in your kitchen, the apple also emits radiation because of its temperature—although we can neither see its radiation nor feel it. When it comes to giving off radiation in this manner, the sun, the apple, and everything else follow the same law—our equation here, which is known as the Stefan-Boltzmann law. In the equation, j^* is energy per unit time (in watts per square meter of surface area) given off by an object, T is its absolute temperature (in Kelvin), and σ is a constant.

At room temperature (let's assume 20°C, or 68°F), the apple is pumping out energy in the form of IR radiation at, according to our equation, 418 watts per square meter. A typical apple might have a surface area of about 0.03 meters squared, and so it would be emitting energy at the not insignificant rate of about 13 watts. But resting comfortably on your kitchen table, the apple is in an environment in which just about everything else is also at room temperature. All of those things are putting off energy at about the same rate as the apple, and thus there is no net energy gain or loss for the apple. It emits radiation, according to the Stefan-Boltzmann law, at pretty much the same rate as it absorbs it, according to the very same law.

But there is at least one thing in your kitchen that is not at room temperature, and that is you. Your core temperature may be 98.6°F, but at your body's surface, things are a bit cooler—about 91.4°F (33°C). If you are wearing clothes, the temperature of the fabric might be something like 28°C. So the Stefan-Boltzmann equation tells us that a clothed human body is putting off radiation at the rate of about 465 watts per square meter. Simultaneously, that same body is absorbing radiation from the surrounding room-temperature environment, as noted above, at 418 watts per square meter. The difference, $465 - 418 = 47$ watts per square meter, is the rate at which your body is radiating IR energy into your kitchen. A typical human body has a surface area of about 2 meters squared, and so it emits radiation at $2 \times 47 = 94$ watts.

To keep things simple, let's round that off to 100 watts. That is the rate at which a typical clothed human body, at rest in a typical room-temperature kitchen, radiates energy. It's about the same as the rate at which heat is given off by a 100-watt incandescent light bulb.* If you've ever touched one that's been on for a while you know that it gets pretty hot—enough to burn your finger. So why isn't your body that hot? Well, the typical human body has about 100 times the surface area of the light bulb, so the 100 watts that are flowing out of the body get spread out over a much larger area than for the light bulb.

Losing energy by radiation, as predicted by the Stefan-Boltzmann law, is perhaps the most important but by no means the only way in which energy

*Incandescent bulbs are notoriously inefficient. A 100-watt tungsten filament bulb has a luminous efficiency of only about 2.6%. That means that less than 3% of the incoming electrical energy is converted to visible light. The rest merely heats up the bulb.

leaves (or enters) the human body. Energy flows due to convection and evaporation are also quite important. Stand in front of a fan, and you will be cooled by convection: the currents of cool air will carry heat away from your body faster than if the air were still. Another mechanism, evaporative cooling, comes from perspiration, which automatically kicks in when the body's core temperature begins to rise as a result of warm weather or physical exertion, for example (see chapter 17). Other energy-flow mechanisms exist as well, but here we focus on radiation, which is not only the most important, in terms of total energy flow, but perhaps also the stealthiest. It goes on day and night, whether you're eating, sleeping, resting, or strenuously exercising. As the apple example above shows, the net energy flow due to radiation depends on two basic things: your surface temperature and the temperature of your surroundings.

The approximately 100 watts of IR radiation emitted by a human body allows us to gain some insight into a variety of things. Let's focus on two that are quite different from each other: climate control in meeting rooms and the basic energy requirements of the human diet.

First, climate control. The more people you pack into a room, the more energy you have to contend with. A party for 30 people in your house will heat things up at the rate of 3,000 watts (30×100), roughly the equivalent of having two electric hair dryers, on high, running full time in your living room. The designers of climate control systems for movie houses, churches, auditoriums, and the like must take this into account. Otherwise, the patrons of these establishments might find themselves uncomfortably warm.

Now, diet. The basal metabolic rate, or BMR, is the minimum rate at which a person consumes energy while resting. This is the energy your body expends in order to keep all its basic systems running, including temperature regulation. It turns out that our 100-watt estimate for human body radiation correlates reasonably well with BMR. The units conversion goes like this: 100 watts = 100 joules per second. Multiply that by the number of seconds in one day (86,400), and you find that the 100 watts of energy flow, over one day, equals 8.64 million joules of total energy. Now convert that to the more familiar unit of food calories: 8.64 million joules equals 2,065 food calories.

Basal metabolic rate, which can be measured experimentally, varies with weight, height, age, and gender. A young man of average height and weight might have a BMR of 1,900 food calories per day—not far at all from our estimate of 2,065 above. Most folks lead lives more active than the basal rate, and so their dietary energy requirements are greater than BMR. But not as much as you might imagine. A person who goes about a daily routine but gets little to no physical exercise might require only about 1.2 times BMR to maintain his or her weight. For our average young man with the BMR of 1,900, that equates to 2,280 food calories per day. Any more than that, and our friend will gain weight—whether he likes it or not.

44. Red Hot

$$E(\lambda,T)=\frac{2hc^2}{\lambda^5}\frac{1}{e^{hc/\lambda kT}-1}$$

The Planck Radiation Law

The Planck radiation law quantifies the energy, E, emitted by an ideal "black body" at an absolute temperature of T across a spectrum of wavelengths, λ. The various constants on the right side of the equation are the speed of light, c; the Planck constant, h; and the Boltzman constant, k. As temperature increases, the equation predicts both that the total amount of energy increases and that the peak energy shifts to lower and lower wavelengths.

Red is a warm color, and blue a cool one, as any fashion designer or interior decorator could tell you. But the connection between color and temperature is much more fundamental than the feelings you get from a new dress or a fresh coat of paint in your dining room.

In our previous story (chapter 43), we tried to convince you that your body is continuously emitting infrared (IR) radiation, and at a not insignificant rate—in fact, at roughly the same rate as a 100-watt light bulb. But you can't see the radiation that your body is emitting, unless you are a boa constrictor or you happen to have one of those fancy IR cameras.

IR cameras measure and create artificial color images of radiation in the IR range of the electromagnetic spectrum. IR waves are invisible to normal cameras, and to the human eye, owing to their longer wavelengths. Visible light has wavelengths of about 400 to 700 nanometers. IR waves extend upwards from 700 nanometers all the way to about 1 millimeter. Now that IR cameras have become more affordable, IR photography has found all kinds of uses. For example, you can hire consultants who will use IR photos to show you where energy is leaking out of your house in the winter through windows, doors, attics, and so on. IR photos of people are fun to look at too. They show that the temperature of the surface of your body is far from uniform. Your nose, ears, and other extremities tend to be quite a bit cooler than the rest of you. What the IR camera is measuring here is the frequency of the IR waves.

The equation in our previous story, the Stefan-Boltzmann law, tells us at what rate objects radiate energy based on their temperature. But what that equation can't tell us is the very information that the IR camera is measuring: the *wavelength* of the radiation being emitted by a body. As its temperature increases, a body gives off more and more energy, and most of that energy is at shorter and shorter wavelengths. This is what our equation here tells us. We'll get back to that in a moment. But first, a quick illustration.

A piece of steel in a blacksmith's shop is perhaps the classic example of what we're talking about. As the blacksmith heats up the steel, it quickly gets so hot that it would give you a nasty burn, should you be so foolish as to try to pick it up with your bare hand. But there is no *visible* evidence that the steel is dangerously hot until the steel has been heated all the way up to about 500°C (932°F). At that temperature, the steel will begin, almost imperceptibly at first, to glow red. As the steel gets hotter and hotter above 500°C, its color changes from dark red to bright red, yellowish-red, and yellow, and finally to a brilliant bluish-white at about 1200°C (2192°F). Your eyes can thus roughly measure the temperature of a piece of hot steel, based on its color, *if* it is hot enough.

But below 500°C, the steel looks, to the unaided human eye, as if it were at room temperature. Not, however, to an IR camera. At room temperature, our piece of steel gives off infrared radiation much like the apple from our previous story. As the steel is heated up, it releases energy more rapidly, and most of that energy is at shorter wavelengths. Eventually those wavelengths get short enough that they are perceptible to the human eye—and we perceive "red hot steel."

Our equation here, called the Planck radiation law, quantifies all of this. There's a lot going on in this equation, but the essence is this: an object at a temperature of T emits radiation with a power density, E (in watts per cubic meter), that lies along a spectrum of wavelengths, λ. The sun is a great example. The energy from the sun that enters the Earth's atmosphere occupies a spectrum that extends from the ultraviolet (the stuff that gives you sunburns), to the visible (the stuff you can see), to the infrared (the stuff we sometimes call "heat waves"). In wavelength terms, ultraviolet waves are shorter than visible, which are shorter than infrared. The actual measured spectrum of sunlight is very accurately modeled using Planck's law, as shown in figure 13.

All objects give off radiation as a function of their temperature. If those objects are hot enough (the sun, a red-hot piece of steel), the radiation is visible

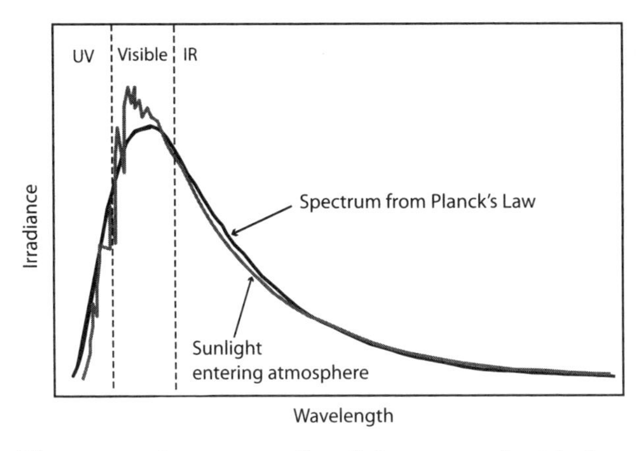

Figure 13. The measured spectrum of sunlight compared with the spectrum as modeled by Planck's law

and our eyes can detect it. For objects that aren't quite so hot (your body, a baked potato), an infrared camera can be used to measure that radiation and convert it into an artificial color pattern we can see.

Boa constrictors and a few other kinds of snakes don't need an IR camera to detect IR waves. But they don't use their eyes either. Just below their eyes, these snakes have separate sense organs called "pit holes" that function as IR detectors. Optically, these organs aren't particularly sophisticated; their ability to focus is quite limited. But they get the job done, functioning in a way that is reminiscent of the old pinhole cameras that some of us (of a certain age) may have constructed when we were kids. Pit holes provide their owners with the equivalent of the night vision goggles soldiers sometimes wear for nighttime operations; snakes thus equipped can hunt in the dark.

Snakes, IR cameras, and radiation from the sun. All of great practical import. But it's well to note that our equation here, relating temperature to energy and wavelength, solved a vexing problem in the realm of classical physics and helped usher in the era of modern physics. Its creator, Max Planck (1858–1947) was a German physicist whose name may or may not ring a bell. Nevertheless, Planck's position in the pantheon of great scientists is more than secure. In 1928, for example, the Max Planck Medal for extraordinary achievement in theoretical physics, which is still awarded today, was presented for the first time. That year, two exceptional scientists were so honored: Max Planck and Albert Einstein.

45. A Bolt from the Blue

$$V = IR$$

Ohm's Law

The voltage, V, in an electrical circuit is equal to the current, I, times the resistance, R. The current is the amount of electrical charge that is flowing in the circuit per unit of time. Current has units of amperes; 1 ampere is 1 coulomb per second. Voltage is the potential difference (measured in volts) between two points in the circuit. Resistance, in ohms, is a measure of the opposition to the flow of current. The equation may be rearranged as $I = V/R$; the current flowing between two points in a circuit is proportional to the potential difference between those points.

Let's face it, the chances of being struck by lightning are pretty slim. But not as slim as you might wish. You are more likely to be hit by lightning than you are to win a lottery jackpot, for example. But that doesn't stop folks from buying lottery tickets. According to the National Weather Service, the odds of an individual in the United States being struck by lightning in a given year are about one in a million.

Lightning is simply a discharge of current caused by an electric imbalance between the clouds and the earth. Lightning is a close (if rather more deadly) cousin of what happens when you walk across a carpet in your socks during the winter and then touch a doorknob. Shuffling across the carpet like that causes a buildup of electric charge on the surface of your body. Touching the metal doorknob provides a path for that current to flow. A shock, and maybe even a visible spark, are the results.

Ohm's law, $V = IR$, our equation here, helps us understand what goes on with static discharges like lightning and doorknobs in winter, even though phenomena such as these are often used to explain conditions under which Ohm's law fails to completely explain things. Confused? Let's take a quick timeout for a few fundamentals, before we get back, in a flash, to our lightning story.

Georg Ohm (1789–1854) was a German physicist. The law that bears his name is one of the most basic and important relationships in the world of physics and electrical engineering. The three variables in Ohm's law stand for voltage (V), current (I), and resistance (R). The electricity that flows through a wire is often compared to the water that's being pumped through a horizontal pipe. The rate at which the water is flowing (in gallons per second, let's say) is analogous to the electrical current, I, in amperes, or coulombs per second, that is flowing through a wire. The roughness of the surface of the pipe creates resistance to the flow of the water, and that resistance is analogous to the electrical resistance, R, inside the wire. (Electrical resistance is measured in ohms in honor of the hero of our story.) Finally, the driving force that causes the water to flow is pressure, which is analogous to the voltage, V, in our wire. The difference in pressure from one end of a water pipe to the other can be calculated from the rate the water is flowing and the internal resistance of the pipe. Likewise, the drop in voltage along the length of a wire can be calculated from Ohm's law, as IR—the current flowing in the wire times the resistance of the wire.

Ohm's law is an empirical relationship, which is to say that it arose from Ohm's observations of various experiments that he carried out. Ohm's law is extremely useful, but it is well known that it has its limitations. The resistance in an electrical wire, for example, changes with the temperature of the wire. The filament in an incandescent light bulb gets extremely hot when a current is passed through it (that's why it glows and gives off light). As the temperature of the filament increases, so does its resistance, and thus the relationship between voltage and current is no longer the linear one predicted by Ohm's law.

Likewise, there are materials whose resistance to the flow of electricity is very high—until those materials are exposed to voltages above a certain threshold. One of those materials is air. And so we're back to our lightning story.

Most of the time, air is a very effective insulator, which just means that air has a very high resistance to the flow of electrical current. The layer of air between you and whatever is behind a "Danger—High Voltage" sign is what is protecting you from electrocution. But if the voltage is high enough, it can cause changes in the air that allow current to be transferred right through the air. In this case, the air is said to "break down" and become ionized, reducing its resistance and allowing for much easier transfer of electrical current. The voltage necessary to cause air to break down and transfer current in this manner is very high, although this property of air is difficult to measure, and it depends on a lot of different variables. Something like 60,000 or 70,000 volts might be necessary to cause electrical current to jump across a one-inch air gap.

When you touch a doorknob in winter, the static electricity discharge can involve voltages as great as 20,000 volts. Because the currents involved are so tiny, these high voltages are nowhere near lethal. (In the case of electric shock, it is current that kills, not voltage.)

Lightning is an altogether different story. Currents of 30,000 amps typically flow in a medium-sized stroke of lightning. Although that stroke only lasts for a tiny fraction of a second (something like 30 millionths of a second), that's enough time to transfer around 500 million joules of energy. This is roughly the amount of energy contained in five gallons of gasoline. Because the energy in lightning is transferred so quickly, the peak power of an average lightning stroke is around one terawatt—a trillion watts! This is about 100 times as great as the power output of the space shuttle at liftoff.

There are lots of things about lightning that are imperfectly understood, including the physics of exactly how a stroke of lightning gets started. But statistics on lightning abound. Lightning occurs in the Earth's atmosphere more than 40 times per second—well over a billion times per year. Most of these strokes go from one cloud to another, although about one stroke in four is from a cloud to the ground. Some places are much more prone to lightning strikes. In the United States, central Florida is "lightning alley." You can look at a map of lightning activity in the United States over the last 60 minutes on the website www.strikestarus.com. On average, about 40 people are killed by lightning strikes in the United States each year. Hundreds of others suffer serious injuries, often resulting in permanent disabilities. The record for being struck by lightning the most times probably goes to Roy Sullivan, who survived seven strikes. No word on whether he ever won the lottery.

46. Like Oil and Water

$$\gamma = \frac{W}{\Delta A}$$

The Surface Tension at the Surface of a Liquid

This equation offers one way to define the surface tension, γ, at the surface of a liquid. W is the amount of energy required to increase the surface area of a liquid by an increment of area equal to ΔA. The units of surface tension, γ, are thus energy per unit area, or J/m^2. Surface tension is often tabulated in units of N/m, which is dimensionally equivalent to J/m^2.

Everyone knows that oil and water don't mix. The old cliché is true—up to a point. But in the petroleum industry, oil is never very far from water, and keeping the two separate is a lot more of a chore than you might have guessed. When crude oil or natural gas is produced from an oil well, water is always present along with the hydrocarbon products, even when the well is hundreds of miles from the nearest ocean, river, or lake. A 2004 report noted that wells in the United States produce, on average, seven barrels of water for each barrel of oil. Separating oil and natural gas from produced water, and then safely disposing of that water, are daily challenges in the petroleum industry.

Sometimes, however, the water that is mixed with oil results not from naturally occurring phenomena, but from an accident of catastrophic proportions. On April 20, 2010, the Deepwater Horizon drilling rig, operating in the Gulf of Mexico about 40 miles from the Louisiana coast, exploded and began to burn. Eleven members of the rig's crew were killed. At the seabed, the well the rig had been drilling began gushing oil and gas into the ocean at a fearful rate. By the time the well was finally capped on July 15, 2010, nearly five million barrels of oil had escaped into the Gulf.

After the accident, armies of workers toiled feverishly to try and staunch the flow of oil from the well while others labored to contain and clean up the oil that had already been spilled. The former task—capping the damaged well—proved spectacularly difficult and took much longer than industry experts had predicted at the outset.

But it is the latter task—containing and cleaning up the spill—that concerns us here. There are only a few basic alternatives for minimizing the environmental impact of an oil spill in the ocean. The oil can be contained and then either collected or burned, or it can be dispersed into the vast expanses of the ocean, where it will eventually biodegrade. Each of these techniques was employed to some extent for the Deepwater Horizon spill. Our focus here is on dispersal.

At various locations on the bottom of the ocean, hydrocarbons naturally leak from the earth into the water. (Once upon a time, there were places on dry land where this occurred as well.) Hydrocarbons that vent naturally into the ocean eventually disperse and biodegrade—but these natural vents are but drops in the bucket, so to speak, compared with the rate at which hydrocarbons were spilled into the ocean in the wake of the Deepwater Horizon disaster.

To accelerate the dispersion and biodegradation process for such a large volume of oil, workers pumped chemicals called dispersants into the ocean down at the seabed adjacent to the leaking well. This approach was controversial for several reasons. First, there was concern about the toxic effects of the dispersant chemicals. Some felt that the medicine (the dispersants) might be worse than the disease (the leaking oil itself) in terms of the long-term effects on the ocean. There was also concern about how well the dispersants would mix with the oil down at the seabed, under the immense pressure of 5,000 feet of water.

The way a dispersant is supposed to work is easy to demonstrate and relatively easy to explain. First, the demonstration. Find two empty jars made of clear glass with lids; jelly jars will do. Fill the jars half full with water, and then add a tablespoon of olive oil to each. The oil, being less dense than water and not very soluble in it, very quickly forms a separate layer on top of the water. Now put the lid on one of the jars and shake it vigorously. The oil disperses itself into the water in small droplets, and the water gets very cloudy. An oil spill in open water such as the Gulf of Mexico is a lot more like the shaken glass than the still one. Currents, waves, and wind all conspire to disperse the oil into the water, making it much more difficult to contain and collect.

Oil in a shaken glass of water, or in the open sea, thus tends to form an emulsion—a dispersion of small droplets of one liquid (oil in this case) in another (water). The two liquids in an emulsion are insoluble. Water and ethanol, in contrast, do not form an emulsion when mixed. Being soluble in one another, they simply form a solution.

Now take the other jar of water and oil (the one you didn't shake). Add a couple drops of dish soap (preferably colorless), close the lid, and shake. Set the two jars side by side where they won't be disturbed. Right from the start, the two oil-water emulsions look quite different. The one with the soap is a bright milky white, with lots of frothy bubbles above the liquid in the half-full jar. The jar with oil and water but no soap is cloudy and yellowish-white. Within a minute or so of shaking the jar with no soap, you will notice that the oil has begun to separate and reform a layer on top of the water. Within hours, the oil and water have almost completely separated, and the water below the oil is nearly as clear as it was to begin with. The soapy oil-water mixture, in contrast, still looks cloudy even weeks after the mixture was created.

Soap belongs to a class of chemicals called surfactants. The two ends of a molecule of soap have very different chemical properties. At one end, a soap molecule is strongly attracted to water. At the other end, the same molecule is strongly attracted to very different types of molecules—such as oil. Add soap to an oil-water mixture, as in our example, and the soap molecules get busy attaching themselves to water molecules at one end and to oil at the other. This has the effect of breaking down the interface between the droplets of oil and the water they are floating around in, in an oil-water emulsion. For the same reason, soap and water are for more effective at lifting grease, dirt, and other things off your skin than water alone is.

That barrier between the oil and water is caused by surface tension; our equation above provides one definition. Surface tension is a familiar phenomenon. Think of an insect such as a water strider skating across the surface of a still pond. Why doesn't it sink? The surface tension of the water essentially forms a "skin" on the liquid surface that allows the bug to walk on water.

The molecules in a liquid are weakly attracted to one another. (If they were much more strongly attracted to one another, our liquid would be a solid; much less strongly, a gas.) In the bulk of the liquid, away from the surface, the weak molecular attractions are more or less the same in all directions. But not at the surface. The molecules at the surface of a liquid are not surrounded on all sides by other liquid molecules. As a result, the surface molecules are more strongly attracted to the other molecules on the surface than the molecules in the bulk of the liquid are attracted to one another. The extra attraction those surface molecules have for one another is what gives the liquid enough tenacity at its surface to allow insects to walk on it, and so forth.

The extra attraction displayed by the surface molecules in a liquid can be measured. Surface tension, γ, is defined as the excess energy at the surface of a liquid, W, per unit area, A. "Excess" in this case means above and beyond the energy associated with the molecules in the bulk of the liquid. Oil cannot break through the surface tension of water, and vice versa. The two liquids thus stubbornly refuse to dissolve in one another, until a surfactant like soap, or one of the chemicals BP used to combat the Deepwater Horizon spill, comes along.

Environmentalists, marine biologists, and others continue to monitor the effects of all of this. We hope that the lessons we learn will never have to be applied to another disaster of this magnitude again.

47. Fish Story

$$l_t = L_\infty\left(1 - e^{-k(t-t_0)}\right)$$

The Von Bertalanffy Equation

At any given time t, a creature will have length l_t. As t increases, the creature gets longer (or taller, depending on your perspective). Length increases rapidly at first, and then slows as it approaches the final length, L_∞. The equation shows that l_t approaches L_∞ asymptotically. The constants k and t_0, as well as L_∞, depend on the creature under consideration.

Boating, while a pleasurable enough leisure-time activity, is not without its hazards. One that you may not have considered is the not insignificant risk, as you zip across certain rivers in the United States, of being struck in the face by a silver carp weighing upwards of 40 pounds that has flung itself high out of the water, having been startled by the sound of your boat's motor.

The silver carp is a type of Asian carp. It is non-native to the United States and is an example of what is known as an invasive species. The silver carp appears to be the only variety of Asian carp with the propensity to leap from the water and into the path of oncoming boaters. That, however, is only the beginning of its destructive behavior.

In the 1970s, several varieties of Asian carp were imported to the United States and stocked in fish farms and wastewater treatment facilities as a means of controlling algae growth. Soon thereafter, the law of unintended consequences took effect, as some of the carp escaped their confines and began to breed at breathtaking rates in a lot of rivers in the central United States. These fish now threaten the Great Lakes, much to the consternation of the $7 billion fishing industry in the world's largest fresh water ecosystem.

When your obnoxious cousin Ralph arrives unannounced for an open-ended visit and begins raiding the fridge with ruthless efficiency, the consequences, while annoying, are hardly life threatening. The same cannot be said for the potential harm of Asian carp to fishing in the Great Lakes. Asian carp feed on plankton and algae, and because they are such voracious eaters, they have the potential to starve out a wide variety of other species because the immature fishes of those other species exist on a diet similar to that of the Asian carp.

Did we mention that these carp are healthy eaters? Asian carp often consume more than 40% of their weight daily. This would be the equivalent of your 200-pound cousin Ralph scarfing down over 80 pounds of food each day. And some carp can eat more than their body weight in a single day, if they can find that much food. Since the carp have no natural enemies here in the United States, their populations continue to increase until they overrun whichever ecosystem they find themselves in. Because the carp typically eat in proportion to their weight, the size of the fish becomes very important.

Our equation here, the von Bertalanffy equation, was developed by Austrian biologist Ludwig von Bertalanffy (1901–72). The equation models the length, l_t, of a creature as a function of time t and the creature's final length, L_∞. In the equation, k and t_0 are constants that depend on the species. The equation predicts rapid growth at first that then slows and approaches final length asymptotically.* Silver carp, for example, which grow to a final length of 78 centimeters (30.7 inches), will have achieved a length of 37 centimeters (14.6 inches) after only one year. That's a good-sized fish, and its prodigious eating habits are no laughing matter for the other species it swims among.

Invasive species arrive by different means. Cousin Ralph may have arrived by Greyhound, while Asian carp, as noted above, were imported in a misguided attempt to control water quality in fisheries and elsewhere. Whether they arrive with or without an invitation, invasive species quickly wear out their welcome and leave us scrambling for solutions to the ecological havoc they wreak. And Asian carp are by no means the only destructive invasive species in the United States.

Kudzu is a large-leaved vining plant that has been called "the vine that ate the South." Anyone who has driven through the southern United States has been struck by the sight of fields, trees, and even small buildings that have been overgrown by lush green kudzu vines. Native to Japan, Korea, and other parts of Asia, the plant grows with frightening swiftness in the moderate climate of the southern United States, where a single vine can grow by up to a foot per day. While the plant has numerous positive qualities, its propensity to overwhelm native species by overgrowing them and depriving them of sunlight

*This sort of behavior—rapid exponential growth that levels off asymptotically—is seen in lots of other things, such as the charging of a capacitor.

has earned the species a notorious place among invasive plant species in the United States. Kudzu stubbornly resists many pesticides. Some success in controlling it has been achieved through grazing by goats and cattle (it makes excellent forage) and by mechanical means (close, frequent mowing).

Zebra mussels are small freshwater creatures that are native to Russia. First detected in the Great Lakes in 1988, they are believed to have arrived via the ballast water of ships traversing the Saint Lawrence Seaway. They infest pipes, pumps, and other equipment in colonies so dense as to choke off or severely limit the operation of the equipment. Like Asian carp, they can outcompete other species for food, and like kudzu, they attach themselves in large numbers on top of, and thus effectively strangle, other species of mussels. As a result, many native mollusk species in the Great Lakes are now endangered. Because zebra mussels are able to live for days out of water, they are often inadvertently transported from one waterway to the next on trailered boats. There is evidence that zebra mussels won't attach themselves to equipment fabricated from copper-nickel alloys. This provides a solution to one of the more important problems that zebra mussels cause, but these alloys are generally much more expensive than the materials they replace.

Controlling Asian carp in U.S. waterways—and in particular, keeping them out of the Great Lakes—has proven to be challenging. The Army Corps of Engineers has deployed electric barriers at various locks entering the Great Lakes. Electric currents in the water agitate fish somewhat analogously to the invisible fences many dog owners use. In a similar fashion, bubble walls can deter passage. Because the fish have gotten into the channels between rivers and lakes, some states, such as Illinois, have taken to poisoning their canals. The biggest stretch poisoned to date was six miles long. Of the fish killed, 90% were common carp.

In the spring of 2010, the Obama administration called for a $78.5 million effort to keep Asian carp out of the Great Lakes. This plan includes a long-range study to determine if the locks should be closed completely. But in the meantime, the locks will not be closed. Instead, lock usage will be limited. The electric barrier will be enhanced and combined with systems of strobes, sounds, and bubbles to improve deterrence. Nets and poisons will be used for those fish that make it past the barrier.

Even as these efforts advance, some have speculated that Asian carp may not be suited for survival in the Great Lakes, where the water is deeper and colder than that of the rivers that feed them. In the end, however, it is probably best not to test the adaptability of such a dangerous species.

48. Making Waves

$$\text{Hull speed} = 1.34\sqrt{\text{Length}}$$

Hull Speed

The hull speed of a boat with a nonplaning hull (such as a sailboat or canoe) is the speed at which the transverse waves the boat is generating—which grow longer as the boat goes faster—achieve a wavelength equal to the length of the boat itself. Hull speed is measured in knots in this equation, while the length of the boat must be expressed in feet.

Let's go boating. Perhaps you're the adventurous type, and you prefer the thrills of a speedboat, which rides up out of the water and skims across its surface almost like a flat stone, expertly hurled almost parallel to the water. Or maybe you're more of a traditionalist and would prefer a sailing vessel, its long, sleek hull gliding silently through the water.

Speedboats and sailboats are members of two distinct classes of watercraft: those with planing hulls and those with nonplaning, or displacement, hulls. A planing hull boat rises up out of the water, like our speedboat, once the boat has reached a certain speed. Displacement hull craft, such as our sailboat, glide through the water, pushing (displacing) the water to either side as they move forward.

For a boat to move through water, the water that's in front of the boat has to end up behind it—just as a car or airplane has to move through air in order to advance. With air, you don't have much of a choice. It's all around, so to get from point A to point B, you have to push your way through the stuff. With water, things are a little different, and it is possible to design things such that a boat can, for the most part, avoid having to push the water out of its way by riding on top of it. Before we get to that, let's look at how boats—the traditional displacement hull kind—move through water.

Humans have been boating since prehistoric times, dating back until probably not long after our ancestors first discovered that wood will float on water. Throughout most of history, boats have been powered by sails or oars—and sometimes both. Steam power ushered in the era of mechanized boats in the 1700s.

Traditional displacement-hulled boats, from canoes to aircraft carriers, rely on buoyancy to remain afloat. In fact, all boats rely solely on buoyancy when they're not moving. The buoyant force pushing a boat up (and thus keeping it from sinking) is equal to the weight of water that the boat displaces. Imagine a rather large toy boat. When you place it in a bathtub, the water level in the tub goes up—that's the water displaced by the boat. The weight of that water is the force pushing up on the bottom of the boat, keeping it from sinking. Boats made from materials that are much denser than water, such as steel, can thus float quite easily, provided that they displace enough water to more than counteract the weight of the boat.

When a displacement-hulled craft, such as a canoe, moves through the water, various things happen to the water around the hull. The subject of hydrodynamics is complex, but in a nutshell, a boat in still, deep water has to overcome two things in order to move forward: frictional effects and the effects of the waves that the boat creates. Water is slippery, and the hulls of ships are smooth. Nonetheless, as a boat moves through water, the water "sticks" to the hull just a bit, creating what is known as a boundary layer. The thicker the boundary layer, the more resistance it offers to the forward motion of the boat. If the surface of a ship were rough like sandpaper, it's easy to imagine that water would stick better to the hull, the boundary layer would be thicker, and the boat would face more resistance to motion.

In addition to frictional effects, a boat also makes waves as it moves through water, and those waves create additional resistance to forward motion. The waves a boat makes are of two basic types: diverging waves and transverse waves, as shown in figure 14. Diverging waves are created at both the bow and stern (front and back) of a boat. These waves travel away from the boat, making an angle with the direction the boat is traveling that varies with the speed of the boat. The familiar wake behind a speedboat is caused largely by these diverging waves. But boats also create a series of transverse waves—waves that are perpendicular to the direction the boat is moving.

The faster a boat is going, the greater the distance between successive transverse waves. The distance between successive waves—the wavelength—increases with the square of the boat's speed. If a boat doubles its speed, this wavelength quadruples. And when a boat moves fast enough that the wavelength of the transverse waves becomes equal to the length of the boat itself, something interesting happens. This situation is depicted by our equation

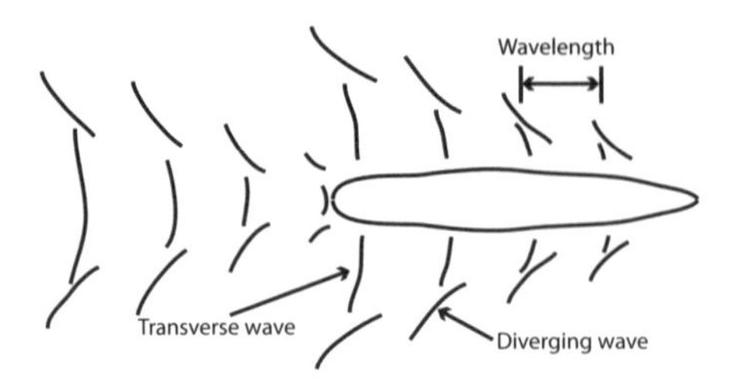

Figure 14. Diverging and transverse waves generated by a boat moving from left to right

here. The "hull speed" of a boat, in knots, is calculated from the square root of the length of the boat measured at the waterline, in feet, times a constant (1.34) to make the units consistent.

What is so special about having a transverse wave whose length is equal to the length of the boat? Well, what this really means is that just as the bow is pushing up one transverse wave at the front of the boat, the peak of the previous wave is passing the back of the boat. You have a wave peak at the very front of the boat, another at the very rear, and in between a giant trough—the valley between the peaks. And that's where the boat sits—right in the middle of the trough. In a sense, the boat is trapped by its own wave.

It was once believed that displacement-hulled craft could not possibly travel any faster than the "hull speed" predicted by our equation. In essence, this would mean that they couldn't go fast enough to climb up out of that trough and outrun the transverse waves they were creating.

For some boats, this is true. The classic example would be the tugboat, a short, squat little boat designed to pull huge vessels in and out of harbors. When tugboats are doing their jobs (tugging other boats), they are traveling very slowly. The shape of a tugboat isn't very hydrodynamic because it doesn't need to be except when it is traveling from one job to the next. As our little friend hustles across the harbor to attach itself to its next customer, its blocky shape kicks up a large bow wave. Because the boat is so short, its hull speed is very low, and the tug appears almost to be sucked down below water level trying to outrace its own wave. Even with its enormous engine, it can't do so.

Other displacement-hulled boats, however, are more than capable of traveling faster than their hull speed. The longer the boat, the more likely this is to be the case. Light boats with long narrow hulls, such as competitive rowing boats, can easily exceed hull speed, in some cases even doubling that speed.

But there's another way to go even faster—the planing hull. With its shallower hull (a displacement hull tends to be much deeper), a planing hull boat isn't subject to hull speed limitations. It simply rides up over the top of its own bow wave, breaking free of most of the hydrodynamic resistance that displacement-hulled craft are forced to deal with. It takes less energy for a hull to plane over the water than it would if the boat were displacing the water at the same speed. Unfortunately, it takes a tremendous amount of energy to make a hull go fast enough to begin planing in the first place. In boating, as in most things, there is no free lunch.

49. A Drop in the Bucket

$$\Delta P = \frac{8\mu LQ}{\pi r^4}$$

The Hagen-Poiseuille Equation

When a fluid flows down a pipe, the pressure the fluid exerts tends to drop from one point to the next, further down the pipe in the direction of the flow. That pressure drop, ΔP, may be calculated from the characteristics of the flow. The equation is for a cylindrical pipe of length L and radius r. The flow rate of the fluid is Q, and its viscosity is μ.

Ours is the age of the all-too-short attention span. Researchers at the University of Pennsylvania found that the impression you make in the first *two seconds* of a job interview may well seal your fate. By comparison, museum-goers these days are much more circumspect. The average patron at an art museum invests about five seconds of her lifetime per canvas before hurrying on to the next masterpiece. And just how many seconds does a TV program have in which to grab the attention of the average couch potato, clicker poised and ready?

Things weren't always this way. In 1927, Thomas Parnell, a professor of physics at the University of Queensland in Australia, began what is essentially a classroom demonstration. That demonstration continues to this day, more than 85 years later. Parnell wanted to demonstrate for his students the extremely slow flow rate of a fluid with very high viscosity. For his fluid, he selected pitch. Pitch is a tarlike substance that can be produced from petroleum products or from wood. If you heat wood without burning it, tar and pitch will eventually drip away from the wood, leaving behind charcoal. Pitch made from petroleum products is, well, it's so black that an absolute, profound darkness is often characterized as "pitch black."

Pitch was traditionally used as a sealant for boats and waterproof containers. To use pitch as a sealant, it first must be heated up so that it can flow. The pitch is then poured or brushed onto the surface to be sealed. Once it has cooled, a very effective and durable watertight seal will have formed.

Professor Parnell started his experiment the same way, by heating a sample of pitch to allow it to flow. He then poured it into a glass funnel with a sealed bottom. Then he waited for the pitch to settle comfortably into the funnel. This took about three years. Parnell, it seems, was not in a hurry.

Finally, the big day arrived, and in 1930, the seal on the bottom of the funnel was cut open and the pitch began to flow out. Which it is still doing today (see fig. 15). Pitch, at room temperature, flows rather slowly, to say the least. But flow it does. Since 1930, exactly eight drops of pitch have flowed out of the funnel and dropped into the glass beaker below. The eighth drop fell in November 2000. Some years earlier, a webcam had been installed to record

Figure 15. The pitch drop experiment at the University of Queensland. *Used by permission of the University of Queensland.*

the event, for no human eyes had ever actually witnessed any of the previous drops detaching from the funnel and plopping into the beaker. In what can only be described as poetic justice, the webcam was on the fritz when the drop finally fell. As of this writing, the ninth drop has yet to fall. It is expected to do so sometime in 2014.

Watching the pitch drop experiment has been described as being like watching paint dry—only more boring. Watching paint dry is hardly boring, however, to a paint chemist. And watching pitch flow out of a glass funnel is hardly dull to someone who specializes in rheology.

Rheology is the study of fluids. Fluids can be characterized on the basis of their flow behavior. Some materials, however, have a bit of a split personality. Such materials behave sometimes like conventional solids and sometimes more like liquids. They are called viscoelastic materials, and pitch is one of them. At room temperature, pitch will shatter if you hit it with a hammer. That's elastic behavior. But put it in a funnel, and at room temperature, it will flow out like a liquid. This is viscous behavior. Except that the "liquid" in this case is really, really, really viscous. Things that move slowly are said to progress "like molasses in January." Might we suggest, for those things that move much, much more slowly, the expression "like pitch in Parnell's funnel?"

Parnell's funnel has attracted the attention of some serious scientists. In a 1984 paper, the viscosity of the pitch in the funnel was estimated using a version of our equation above. Poiseuille's law (or the Hagen-Poiseuille equation), as it is known, relates ΔP, the pressure drop across a fluid flowing through a cylindrical pipe, to the length of the pipe, L; its radius, r; the flow rate, Q (in liters per minute, for example); and the viscosity of the fluid, μ. After making corrections for the shape of the funnel, the weight of the pitch, and the changing temperature of the environment (for decades, the experiment was housed in a building that was not air-conditioned), the authors concluded that the average viscosity of the pitch was something like 2.3×10^8 pascal-seconds (Pa-s). The viscosity of water at room temperature is 1×10^{-3} Pa-s, which is equal to 0.001 Pa-s. The viscosities of thicker fluids at room temperature include that of honey (2–10 Pa-s), creamy peanut butter (250–350 Pa-s), and lard (1,000–2,000 Pa-s). Lard, at room temperature, is thus at least a million times more viscous than water—and the pitch in the funnel is something like 100,000 times as viscous as lard.

Thomas Parnell died in 1948, about a year and a half after the second drop fell. In 1961, a year before the fourth drop fell, care of the pitch drop experiment was placed in the capable hands of Professor John Mainstone, who passed away in August 2013. For more than 50 years, he measured historical events in terms of which drop they preceded or succeeded. Humans first walked on the moon in 1969, a year before the fifth drop. The seventh drop fell in 1988. A year later the Iron Curtain followed suit. As a timepiece, the pitch drop experiment might be found rather lacking. But it's somehow comforting knowing that, for the better part of a century, a simple classroom experiment has been educating students and reminding us all that things will not always be the way that they have always been.

50. Fracking Unbelievable

$$p_b = T - \sigma_H + 3\sigma_h$$

Breakdown Pressure to Initiate Rock Fracture

An equation from the rock mechanics of hydraulic fracturing, for the case of a vertical borehole in a nonporous, impermeable formation. The breakdown pressure necessary to initiate rock fracture, p_b, is related to the tensile strength of the rock, T. The other terms on the right hand side involve σ_H and σ_h, which are the principal stresses in the horizontal plane of the wellbore. By convention, σ_H is larger than σ_h.

On November 11, 2011, an earthquake of magnitude 5.6 on the Richter scale rocked central Oklahoma. It was the strongest quake in the state in almost 60 years. Coincidentally—or not—central Oklahoma is also home to quite a bit of oil- and gas-well drilling activity. Is there a connection here?

The production of crude oil and natural gas from wells seems, to many folks, like a pretty low-tech operation. It brings to mind grainy old black-and-white movies of workers in hard hats, covered in the black slime that's gushing out from a well beneath a quaint wooden derrick, while safely off in the distance, the workers' bosses, decked out in fancy suits, pop corks on champagne bottles to celebrate yet another success.

Modern oil and gas production bears very little resemblance to such antiquated stereotypes. Virtually every aspect of drilling wells and producing hydrocarbons has gone high tech. Perhaps nothing has revolutionized the industry more than two staples of the modern petroleum industry: directional drilling and hydraulic fracturing. Directional drilling sounds impossible. You start by drilling straight down into the earth, and then, somehow, you gradually turn the corner and, voilà, now you're drilling sideways, or horizontally with respect to the surface of the Earth.

In fact, however, what's really difficult is to drill a truly straight hole, as any weekend warrior with a hand-held electric drill could tell you. In the early days of the petroleum industry, when wells supposedly went "straight down," they actually zigged and zagged all over the place. There was no way to know where the bottom of the hole was located at any given point in the process. And even if the drill team had known where the bottom of the hole was, they had very little ability to control where they were going with it. Nowadays, well drilling is controlled so precisely that a well can be drilled vertically for several miles, and horizontally a similar distance, such that the end of the hole hits a target not much larger than the front door of your house.

Precisely controlled directional drilling is important because the well needs to pass through the right geological formations—the places where all the hydrocarbons are located. These have previously been pinpointed by the geologists working on the project. Just how they figure out where all of the oil and gas is located is another story for another day. But once the geologists have done their work, it's up to the drilling engineers to create a well that tunnels through the subterranean real estate most likely to yield the richest harvest of hydrocarbons.

But just drilling a well that follows the best path is not enough. These days, as often as not, the hydrocarbon resources are locked up inside underground formations. That means that very little production is realized simply by drilling a well. This is where the second technology that has revolutionized the oil and gas industry comes in: hydraulic fracturing—commonly known as fracking.

Hydraulic fracturing refers to the cracking open of rock formations by the action of liquids under high pressure. Rocks have been getting cracked open due to hydraulic fracturing pretty much as long as the planet Earth has been around. Natural hydraulic fracturing is billions of years old. You can see evidence of this every time you drive along a highway that has been cut through a rock formation. The veins that extend through the rock walls on either side of the road are often caused by natural hydraulic fracturing. Dig up a stone in your garden, and there's a reasonable chance that it too will contain veins induced by hydraulic fracturing.

Hydraulic fracturing of the human-made variety dates back only to 1947. The basic idea is relatively simple. Once a well has been drilled, liquids are pumped down the hole at very high pressures—15,000 pounds per square inch is not unusual these days. This fractures the rock, creating fissures in the formations that the well passes through and making it much easier for hydrocarbons to find their way through the formation via these fissures, into the well, and up to the surface. Our equation here comes from the field of rock mechanics. The breakdown pressure, p_b, induced by the hydraulic fracturing pumps, that is required to initiate rock fracture is related to the tensile strength of the rock, T, and to stresses induced in the rock by gravity and other forces, σ_H and σ_h.

While the idea of hydraulic fracturing is relatively simple, the details are quite a bit more complicated. Most hydraulic fracturing takes place in wells that are lined with a steel casing. That casing has to be perforated in order for the high-pressure liquid to get out into the formation and do its job. The fracturing liquids themselves are complex—and controversial. The main ingredients are water and sand. Sand (and similar materials) serve as "proppants." Once a crack forms in the rock, the sand flows in and props open the crack, preventing it from collapsing shut when the fracking pressure is reduced. Lots of other chemicals can be added to the fracturing liquids depending on the particulars of a given well. Most wells are fractured a little bit at a time. The very end of the well (the last part to be drilled) is fractured first, then that section of the well is plugged off, and the next section is fractured, and so on. This process could be repeated dozens of times for a single well. At the end, all of the plugs have to be removed before the well can begin to produce. Certain types of formations, including the shale rock often in the news these days, can only be economically exploited with the aid of hydraulic fracturing.

Unless you've been living under a rock, you're probably aware that as hydraulic fracturing has become more popular, it has also become more controversial. The potential environmental impacts of hydraulic fracturing range from groundwater contamination to greenhouse gas emissions to the potential to stimulate seismic activity (earthquakes). Public opinion on hydraulic fracturing is somewhat fractured, to say the least. It ranges from those who believe the process is completely safe and should not be subject to any regulation whatsoever, to those who feel it is so inherently dangerous that it should be forbidden. (In 2012, Vermont became the first state to ban hydraulic fracturing within its borders.) The truth, as usual, is to be found somewhere in between these two extremes. As for the Oklahoma earthquake with which we began this story, it almost certainly had nothing to do with hydraulic fracturing (which is not to say that fracking is incapable of precipitating seismic activity).

Our thirst for petroleum products seems to know no bounds. Hydraulic fracturing gives us a way to quench that thirst, providing access to resources once thought to be unrecoverable. Until the demand side of this equation changes, the supply side, including hydraulic fracturing, is unlikely to, either.

51. Take Two Aspirins and Call Me in the Morning

$$Z = \frac{\bar{X} - \bar{Y}}{\sqrt{\frac{\sigma_1^2}{m} + \frac{\sigma_2^2}{n}}}$$

Determining Statistical Significance

One statistic, Z, sometimes employed to evaluate the significance of the difference between the mean values, $\bar{X}$ and $\bar{Y}$, of two sample populations. The larger the value of Z, the more likely it is that the difference between $\bar{X}$ and $\bar{Y}$ is "statistically significant." The standard deviations associated with $\bar{X}$ and $\bar{Y}$ are σ_1 and σ_2, while m and n are the number of samples from which $\bar{X}$ and $\bar{Y}$ were calculated.

Modern medicine has become so powerful that it's easy to forget that once upon a time, doctors could do little beyond consoling their dying patients while trying to make the patients' last hours as comfortable as possible. Lacking the tools to effectively treat a whole host of ailments, physicians of old would sometimes prescribe treatments that they, the physicians, knew had no curative powers. But that's not what they told their patients. The patients were led to believe that these were powerful therapies capable of restoring their health. And sometimes, they did get better. Call it the power of positive thinking, dumb luck, or, somewhat more scientifically, the placebo effect.

Placebo is Latin for "I will please." It appears in a fourth-century Latin version of the Bible, in Psalm 116:9: *Placebo Domino in regione vivorum*, "I will please the Lord in the land of the living." Over time, however, the word grew to have a somewhat pejorative meaning, to connote one who used deception to please others as a means to an end. A character in Chaucer's fourteenth-century *Canterbury Tales*, called Placebo, is a wicked flatterer.

The first medical use of the term "placebo" dates to 1785. In 1811, *Quincy's Lexicon-Medicum* provided this definition of "placebo": "An epithet given to any medicine adapted more to please than to benefit the patient." In 1807,

Thomas Jefferson wrote: "One of the most successful physicians I have ever known has assured me that he used more bread pills, drops of colored water, and powders of hickory ashes, than of all other medicines put together." Well into the twentieth century, placebos played an important role in mainstream medicine.

Few physicians thought twice about this. Given the limited range of legitimate therapies at their disposal, they were essentially powerless to truly treat a whole host of ailments. Prescribing a placebo, it seemed, was better than doing nothing at all. Medically, it couldn't hurt, and most ethical standards of the day saw nothing wrong with the practice. What's more, many physicians believed that the less intelligent the patient, the more valuable and necessary placebos became.

In more recent years, the placebo has played a vital role in the development of new therapies. Nowadays, just about everyone takes for granted that the efficacy of a new medical treatment can and should be tested experimentally and evaluated statistically. The goal is to find out whether two sets of data are significantly different from one another. Our equation above shows one way, among many, for determining statistical significance. For example, imagine that you have a fish farm, and you want to compare two different feeding regimens to see which yields the largest fish. You set up an experiment in two different fish ponds, being careful to control a whole host of variables (sunlight exposure, pond depth, water quality, feeding times, and so on). You then feed the fish in the first pond with one regimen, and those in the second with the other. After so many days, you harvest the fish and weigh them.

The average weights of the fish in the two ponds, per our equation above, are $\bar{X}$ and $\bar{Y}$. Obviously, the greater the difference in $\bar{X}$ and $\bar{Y}$, the more likely it is that the choice of feeding regimen made a significant difference in the weights of the fish. But our equation takes more into account than just the difference in average weight. In the equation, the number of fish harvested from the first pond is m and from the second it is n. The standard deviations of the weights of the fish from the two ponds are σ_1 and σ_2. All of this allows us to calculate Z, which is a standard statistical quantity whose value can be compared with tables to determine whether the difference between $\bar{X}$ and $\bar{Y}$ is large enough to be considered "statistically significant." This and other similar approaches are used throughout science and engineering and are common in clinical medical trials. A graphic example is shown in figure 16.

Well into the 1900s, however, the effectiveness of a proposed new medical treatment was typically judged not by a clinical trial, but somewhat arbitrarily

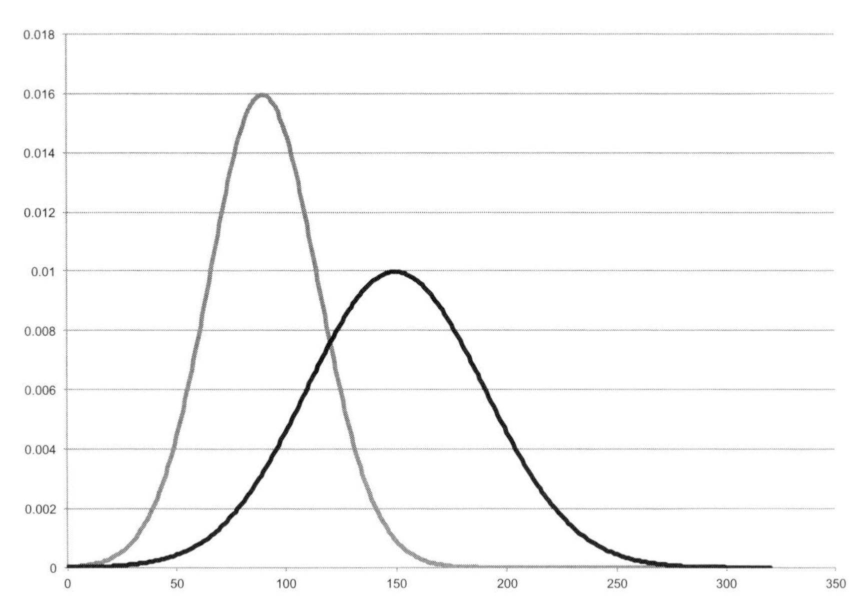

Figure 16. The curves on this graph represent two sets of normally distributed data. The gray curve has a mean of 90 and a standard deviation of 25. The black curve has a mean of 150 and a standard deviation of 40. Each curve was generated from the results of 30 tests. The two sets of data are significantly different.

through the opinions of experts. The first recorded use of placebos as a control in a clinical trial dates to 1801, when the physician John Haygarth compared the application of wooden rods to the skin of patients with the application of metallic rods. In those days, certain metallic rods (known as "Perkins tractors") were believed to have curative powers due to their magnetic properties. Wood has no such properties, yet Haygarth found that patients were just as likely to experience relief from a variety of symptoms after treatment with wooden rods (which they were told were metal) as with authentic Perkins tractors.

In 1938, placebos were used as controls in trials involving the use of vaccines designed to prevent the common cold. Two groups of students were given either the new vaccine or a placebo. Students were not told which treatment they had received, and the treatments were administered identically. A significant percentage of those given the new vaccine reported no colds throughout the trial period. Does this mean the new vaccine was effective? No, because a similar percentage of those receiving the placebo *also* reported no colds during the trial period. Thus, the results of the trial were negative.

The most remarkable aspect of this investigation, the clinicians reported, was that the placebo group showed as much improvement as had previous groups that received vaccines in earlier, uncontrolled trials (trials with no placebo).

This and similar results in other trials led to the realization that administering a placebo during a clinical trial is not the same as doing nothing. There is, in fact, a "placebo effect."

Realizing that there is a placebo effect is one thing. Defining it is another. One definition of the placebo effect is "the difference in outcome between a placebo treated group and an untreated control group in an unbiased experiment." This sounds good in theory, but in practice it is difficult to render such an experiment truly unbiased because you cannot, by definition in this case, disguise who is receiving the (placebo) treatment and who isn't.

A patient's state of mind has a powerful influence on his or her condition. Consider a 1987 study in which 200 patients were divided into four groups. The first group was given a treatment and a "positive" consultation from the physician (the physician spoke in positive terms about the patient's condition and prognosis). The second group was given the same treatment as the first along with a "negative" consultation. The third group received no treatment and a positive consultation, while the fourth received no treatment and a negative consultation. The results of the study showed that the patients who received a positive consultation were far more likely to get better, regardless of whether they received the treatment. Medicine has come a long way from the days in which doctors could do little more than deceive their patients with placebos. But never underestimate the power of positive thinking.

52. The World's Most Famous Equation

$$E = mc^2$$

The Equivalence between Mass and Energy, as Formulated by Albert Einstein

The nuclear energy contained in a quantity of matter of mass m is equal to m times the speed of light, c, squared. The speed of light (see chapter 34) is equal to 299,792,458 meters per second, or approximately 3×10^8 m/s. The SI unit of energy is the joule, which is equivalent to $\text{kg}(\text{m}^2)/\text{s}^2$.

The year 1905 was a very good one for Albert Einstein. As good years go in the world of science, they don't get much better than this one. In 1905, Einstein was a 26-year-old assistant patent examiner in the Swiss patent office, a job he had taken because he had been unable to find work as a college professor of physics. But when he wasn't evaluating folks' inventions for patentability, Einstein spent his spare time working out various scientific puzzles. His science laboratory, essentially, was located in between his ears. And what a laboratory it was.

In 1905, Einstein published four papers in the journal *Annals of Physics*, any one of which would have represented the pinnacle of practically any other scientific career. One of those papers, on a particle interpretation of light, predicted the existence of the photon—the fundamental particle of light. One of the applications of this idea is called the photoelectric effect, to which we shall return presently. Another of Einstein's 1905 papers, on Brownian motion, confirmed the existence of atoms and molecules—which had only been theorized up to that point. A third article laid out the special theory of relativity, forever changing our notions of space and time. A fourth posited an equivalence between mass and energy and included the most famous equation of them all, $E = mc^2$. Not long after these papers appeared in print, Einstein began to be offered the professorships he so obviously merited. Today, we refer to the year 1905 in Einstein's career as the *Annus Mirabilis*, Latin for "extraordinary year."

Thus began the career of perhaps the best-known scientist of all time. Einstein kept working and creating for another 50 years after the *Annus Mirabilis*, before passing away in 1955. More than 50 years after his death, his image—with all that wild hair, the bushy mustache, and the quizzical expression—is as familiar as ever. As dead celebrities go, his is one of the most valuable franchises, right up there with Elvis and Marilyn Monroe.

If you ask regular folks what Albert Einstein did that was so great, the two most common answers would likely be "relativity" and "$E=mc^2$." The greatest achievements of the greatest scientists, as everyone knows, are rewarded with the Nobel Prize. Established by Alfred Nobel and awarded annually since 1901, the Nobel Prizes are given in physics, chemistry, and physiology or medicine. (Nobel Prizes are also awarded in literature and for peace.)

Albert Einstein did receive a Nobel Prize for physics, but it had nothing to do with either relativity or $E=mc^2$. And Einstein had to be nominated for the prize 11 different times before he finally won, in 1921. The citation on Einstein's Nobel Prize read: "For his services to Theoretical Physics, and especially for his discovery of the law of the photoelectric effect." Awarding the Nobel Prize in physics to Albert Einstein for the photoelectric effect would have been a little bit like putting Babe Ruth in the Baseball Hall of Fame because he was a great pitcher. Babe Ruth *was* a great pitcher, as most students of baseball history could tell you, but he's in the Hall of Fame primarily because he was a great hitter.

Einstein's greatest contributions could probably be summed up in one word: relativity. But our story here relates to mass-energy equivalence, because of that famous equation. The two concepts are, in fact, related.

Einstein was not the first person to theorize an equivalence between mass and energy, but he was the first to quantify it—to set it down in equation form. In $E=mc^2$, E is energy, m is mass, and c is the speed of light. The paper in which Einstein proposed $E=mc^2$ is entitled "Does the Inertia of a Body Depend upon Its Energy Content?" This paper is often described as an afterthought to a paper Einstein wrote a few months earlier ("On the Electrodynamics of Moving Bodies"), in which he sets forth the special theory of relativity.

Because the speed of light is so large (about 300 million meters per second, or 700 million miles per hour), even a very small amount of mass is equivalent to an enormous amount of energy. $E=mc^2$ is often used to explain why nuclear weapons are so powerful. The atomic bomb that was dropped on Nagasaki at the end of World War II was equivalent to about 21,000 tons of TNT, a conventional (non-nuclear) explosive. The Nagasaki bomb contained just over six kilograms (about 13 pounds) of plutonium. Roughly one kilogram's worth of those plutonium atoms split apart into lighter atoms, resulting in an unimaginable release of energy. The difference in mass between the original one kilogram of plutonium and the lighter atoms created from the plutonium in the explosion was very close to one gram. That single gram of mass was carried away in the radiation, thermal energy, and blast energy of the explosion. The nuclear energy in one gram of plutonium is thus equivalent to the energy given off in an explosion of about 21,000 tons of TNT.

But $E=mc^2$ is true in general, for all kinds of energy, not just for nuclear energy. It's just that for "conventional" forms of energy, changes in energy are so small that the amount of mass they represent is, generally speaking, pretty much negligible. To convince yourself of that, rearrange the equation to read $E/c^2=m$. Dividing a relatively small amount of energy by c^2, an enormous number, yields a result for m that is very close to zero. When, for example, you heat up water on your stove, the thermal energy you are adding to the water actually increases the water's mass. But not by so much that you'd notice. To heat a gallon of water from room temperature to the boiling point does not add more than a few trillionths of a gram of mass to the water. The reverse of this process—cooling the water from 212°F to room temperature—reduces the mass of the water by the same amount.

In the most famous equation of them all, Albert Einstein showed us that mass and energy are two sides of the same coin, forever linked by a constant: the speed of light squared.

Bibliography

Preface

Adler, Mortimer J., and Charles Van Doren. *How to Read a Book*. Revised edition. Touchstone, 1999.
Thompson, S. P. *The Life of William Thomson, Baron Kelvin of Largs*. Macmillan, 1910.

1. As the Earth Draws the Apple

White, Michael. *Isaac Newton: The Last Sorcerer*. Basic Books, 1999.

2. And All the Children Are Above Average

Devore, Jay L. *Probability and Statistics for Engineering and the Sciences*. 3rd edition. Brooks/Cole Publishing, 1991.
Lawrence, Mark. "Statistics, Part 1: Average and Standard Deviation." investing .calsci.com/statistics.html (accessed September 24, 2012).

3. The Lady with the Mystic Smile

Livio, Mario. "The Golden Ratio and Aesthetics." *Plus Magazine*, no. 22, 2002. http://plus.maths.org/issue22/features/golden.
Markowsky, George. "Misconceptions about the Golden Ratio." *College Mathematics Journal* 23, no. 1 (January 1992).
Walser, Hans. *The Golden Section*. Mathematical Society of America, 2001.

4. The Heart Has Its Reasons

Mitchell, G., and J. D. Skinner. "An Allometric Examination of the Giraffe Cardiovascular System." *Comparative Biochemistry and Physiology, Part A*, 154 (2009): 523–29.
Simmons, Robert, and Lee Scheepers. "Winning by a Neck: Sexual Selection in the Evolution of the Giraffe." *American Naturalist* 148 (1996): 771–86.

5. AC/DC

King, Gilbert. "Edison vs. Westinghouse: A Shocking Rivalry." October 11, 2011. http://blogs.smithsonianmag.com/history/2011/10/edison-vs-westinghouse-a-shocking-rivalry (accessed August 25, 2012).

6. The Doppler Effect

Eden, Alec. *The Search for Christian Doppler.* Springer-Verlag, 1992.

7. Do I Look Fat in These Jeans?

Centers for Disease Control and Prevention. U.S. National Health and Nutrition Examination Survey. www.cdc.gov/nchs/nhanes.htm (accessed September 24, 2012).

Eknoyan, Garabed. "Adolphe Quetelet (1796–1874): The Average Man and Indices of Obesity." *Nephrology Dialysis Transplantation* 23, no. 1 (2008): 47–51.

8. Zeros and Ones

Smith, Elizabeth. "On the Shoulders of Giants: From Boole to Shannon to Taube; The Origins and Development of Computerized Information from the Mid-19th Century to the Present." *Information Technology and Libraries* 12, no. 2 (June 1993): 217–26.

9. Tsunami

Back, Alexandra. "Tsunamis: How They Form." *Australian Geographic,* March 18, 2011. www.australiangeographic.com.au/journal/facts-and-figures-how-tsunamis-form.htm.

John, James E. A., and William L. Haberman. *Introduction to Fluid Mechanics.* 3rd edition. Prentice Hall, 1988.

10. When the Chips Are Down

Moore, Gordon. "Cramming More Components onto Integrated Circuits." *Electronics* 38, no. 8 (April 19, 1965): 4–7.

———. "Progress in Digital Integrated Electronics." *Technical Digest,* IEEE International Electron Devices Meeting 21 (1975): 11–13.

Yang, Dori Jones. "Gordon Moore Is Still Chipping Away." *U.S. News & World Report,* July 20, 2000. www.usnews.com/usnews/biztech/articles/000710/archive_015221.htm.

11. A Stretch of the Imagination

Chapman, Allan. "England's Leonardo: Robert Hooke (1635–1703) and the Art of Experiment in Restoration England." *Proceedings of the Royal Institution of Great Britain* 67 (1996): 239–75. http://home.clara.net/rod.beavon/leonardo.htm.

12. Woodstock Nation

Feldman, David. *Imponderables: The Solution to Mysteries of Everyday Life*. Pp. 250–54. William Morrow, 1987.

13. What Is π?

Bell, E. T. *Men of Mathematics*. Pp. 28–34. Simon and Schuster, 1937.

14. No Sweat

"Cooling of the Human Body." Georgia State University, HyperPhysics website. http://hyperphysics.phy-astr.gsu.edu/hbase/thermo/coobod.html#c1 (accessed September 25, 2012).

Guyton, Arthur C., and John E. Hall. *Textbook of Medical Physiology*. Pp. 889–904. 11th edition. Elsevier Saunders, 2006.

15. Road Range

Anderson, Curtis D., and Judy Anderson. *Electric and Hybrid Cars: A History*. McFarland, 2004.

16. The Bends

McCullough, David. *The Great Bridge*. Simon and Schuster, 1972.

17. It's Not the Heat, It's the Humidity

Middleton, W. E. Knowles. *A History of the Thermometer and Its Uses in Meteorology*. Johns Hopkins University Press, 2002.

18. The World's Most Beautiful Equation

Nahin, Paul J. *Dr. Euler's Fabulous Formula: Cures Many Mathematical Ills*. Princeton University Press, 2006.

Sandifer, Ed. "How Euler Did It: Euler's Greatest Hits." Mathematical Association of America website. www.maa.org/editorial/euler/How Euler Did It 40 Greatest Hits.pdf (accessed September 25, 2012).

19. Breaking the Law

Hunt, Bruce J. *Pursuing Power and Light: Technology and Physics from James Watt to Albert Einstein*. Johns Hopkins University Press. 2010.

Schirber, Michael. "Harsh Light Shines on Free Energy." *Physics World* 20 (August 2007): 9.

20. The Mars Curse

O'Neill, Ian. "The Mars Curse." March 22, 2008. www.universetoday.com/13267/the-mars-curse-why-have-so-many-missions-failed.
Stephenson, Arthur G., et al. "Mars Climate Orbiter Mishap Investigation Board Phase I Report." NASA, November 10, 1999.

21. Eureka!

Bell, E. T. *Men of Mathematics.* Pp. 28–34. Simon and Schuster, 1937.

22. A Penny Saved . . .

O'Connor, J. J., and E. F. Robertson. "History Topic: The Number *e*." www-history.mcs.st-and.ac.uk/PrintHT/e.html (accessed December 7, 2013).

23. If I Only Had a Brain

Maor, Eli. *The Pythagorean Theorem: A 4,000-Year History.* Princeton University Press, 2007.

24. Because It Was There

Christiaens, Griet. "The Prince of Amateurs of Mathematics." http://mathsforeurope.digibel.be/pierredefermat.html (accessed September 25, 2012).
Wiles, Andrew. "Modular Elliptic Curves and Fermat's Last Theorem." *Annals of Mathematics* 141, no. 3 (1995): 443–551.

25. Four Eyes

Ilardi, Vincent. *Renaissance Vision from Spectacles to Telescopes.* American Philosophical Society, 2007.

26. Bee Sting

Lonsdorf, Eric, et al., "Modelling Pollination Services across Agricultural Landscapes." *Annals of Botany*, 2009, doi: 10.1093/aob/mcp069; first published online March 26, 2009.
Madrigal, Alexis. "Bee Colony Collapse May Have Several Causes." www.wired.com/wiredscience/2010/01/colony-collapse-lives (accessed September 3, 2012).

27. Here Comes the Sun

Boyles, Sally. "High-SPF Sunscreens: Are They Better?" www.webmd.com/skin-problems-and-treatments/features/high-spf-sunscreens-are-they-better (accessed September 5, 2012).

Brannon, Heather. "What Is SPF?" http://dermatology.about.com/cs/skincareproducts/a/spf.htm (accessed August 25, 2012).

Derrick, Julyne. "Top 10 Sunscreens." http://beauty.about.com/od/sunscreel/tp/sunscreenstop.htm (accessed September 5, 2012).

28. A Leg to Stand On

LaBarbera, Michael L. "The Biology of B-Movie Monsters." http://fathom.lib.uchicago.edu/2/21701757 (accessed September 5, 2012).

29. Love Is a Roller Coaster

Serway, Raymond, and Robert Beichner. *Physics for Scientists and Engineers*. 5th edition. Pp. 979–94. Saunders College Publishing, 2000.

30. Loss Factor

Feynman, Richard. *What Do You Care What Other People Think? Further Adventures of a Curious Character*. W. W. Norton, 2001.

Gebhardt, Chris, and Chris Bergin. "STS-51L and STS-107—*Challenger* and *Columbia*: A Legacy Honored." February 1, 2010. www.nasaspaceflight.com/2010/02/sts-51l-sts-107-challenger-columbia-legacy-honored (accessed September 5, 2012).

31. A Slippery Slope

"Friction and Coefficients of Friction." www.engineeringtoolbox.com/friction-coefficients-d_778.html (accessed September 25, 2012).

"Introduction to Tribology: Friction." http://depts.washington.edu/nanolab/ChemE554/Summaries ChemE 554/Introduction Tribology.htm (accessed September 5, 2012).

Serway, Raymond, and Robert Beichner. *Physics for Scientists and Engineers*. 5th edition. Pp. 131–37. Saunders College Publishing, 2000.

32. Transformers

Devlin, Keith "The Maths behind MP3." *The Guardian*, April 3, 2002. www.guardian.co.uk/technology/2002/apr/04/internetnews.maths.

Fourier, Joseph. "Remarques générales sur les températures du globe terrestre et des espaces planétaires." *Annales de Chimie et de Physique* (Paris), 2nd ser., 27 (1824): 136–67.

Maor, Eli. *Trigonometric Delights*. Pp. 198–210. Princeton University Press, 1998.

33. A House of Cards

Darby, Mary. "In Ponzi We Trust." *Smithsonian Magazine*, December 1998. www.smithsonianmag.com/people-places/In-Ponzi-We-Trust.html.

34. Let There Be Light

Boyer, Carl. "Early Estimates of the Velocity of Light." *Isis* 33, no. 1 (March 1941): 24–40.

35. Smarty Pants

Martin, O. "Psychological Measurement from Binet to Thurstone (1900–1930)." *Revue de Synthese* 4 (1997): 457–93.
"The SAT." http://professionals.collegeboard.com/testing/sat-reasoning (accessed September 25, 2012).

36. As Old as the Hills

Dalrymple, G. Brent. *The Age of the Earth*. Stanford University Press, 1991.

37. Can You Hear Me Now?

"Reverberation Time." http://hyperphysics.phy-astr.gsu.edu/hbase/acoustic/revtim.html (accessed September 25, 2012).
Sabine, Wallace C. *Collected Papers on Acoustics*. 1922. Reprint: Dover, 1964.

38. Decay Heat

Decay Heat Power in Light Water Reactors. American Nuclear Society, ANSI/ANS-5.1-2005.
Nusbaumer, Olivier. "Decay Heat in Nuclear Reactors." http://decay-heat.tripod.com (accessed September 5, 2012).

39. Zero, One, Infinity

Drake, Frank. "The Drake Equation Revisited: Part 1." *Astrobiology Magazine*. September 29, 2003. www.astrobio.net/index.php?option=com_retrospection&task=detail&id=610.

40. Terminal Velocity

"Red Bull Stratos Fact Sheet." http://media.marketwire.com/attachments/201002/2757_RedBullStratos-SupersonicFactSheet-long.pdf (accessed September 6, 2012).

"Supersonic Parachuting: Red Bull Stratos vs Le Grand Saut." April 10, 2010. www.whitelabelspace.com/2010/04/supersonic-parachuting-red-bull-stratos.html (accessed September 6, 2012).

41. Water, Water, Everywhere

"The World Factbook: Bangladesh." www.cia.gov/library/publications/the-world-factbook/geos/bg.html (accessed September 6, 2012).

Yavuz, C. T., et al. "Low-Field Magnetic Separation of Monodisperse Fe_3O_4 Nanocrystals." *Science*, November 10, 2006, pp. 964–67.

42. Dog Days

Medawar, P. B. *An Unsolved Problem in Biology*. H. K. Lewis, 1952.

Reznick, D. N., et al. "Effect of Extrinsic Mortality on the Evolution of Senescence in Guppies." *Nature* 431 (2004): 1095–99.

43. Body Heat

McCardle, William, Frank Katch, and Victor Katch. *Exercise Physiology: Energy, Nutrition, and Human Performance*. 5th edition. Lippincott, Williams, and Wilkins, 2001.

Power, Michael, and Jay Schulkin. *The Evolution of Obesity*. Johns Hopkins University Press, 2009.

44. Red Hot

Britton, Erin. "Planck's Law." December 28, 2008. http://physics-history.suite101.com/article.cfm/plancks_law (accessed September 18, 2012).

Planck, Max. *The Theory of Heat Radiation*. Translated by M. Masius. 2nd edition. P. Blakiston's Son, 1914.

45. A Bolt from the Blue

Serway, Raymond, and Robert Beichner. *Physics for Scientists and Engineers*. 5th edition. Pp. 844–52. Saunders College Publishing, 2000.

Uman, Martin A. *All about Lightning*. Dover, 1987.

"United States Lightning Activity, Last 60 Minutes," www.strikestarus.com (accessed September 18, 2012).

46. Like Oil and Water

Cressey, Daniel. "The Science of Dispersants: Massive Use of Surfactant Chemicals Turns Gulf of Mexico into a Giant Experiment." *Nature*, May 12, 2010. www.nature.com/news/2010/100512/full/news.2010.237.html.

Veil, John A., et al. *A White Paper Describing Produced Water from Production of Crude Oil, Natural Gas, and Coal Bed Methane*. Argonne National Laboratory, for the U.S. Department of Energy National Energy Technology Laboratory. January 2004.

47. Fish Story

Garvey, J. E., K. L. DeGrandchamp, and C. J. Williamson. *Life History Attributes of Asian Carps in the Upper Mississippi River System*. ANSRP Technical Notes Collection, ERDC/TN ANSRP-07-1. U.S. Army Corps of Engineer Research and Development Center, Vicksburg, MS, May 2007. www.dtic.mil/cgi-bin/GetTRDoc?Location=U2&doc=GetTRDoc.pdf&AD=ADA468471 (accessed September 18, 2012).

"The Von Bertalanffy Growth Equation." www.pisces-conservation.com/growthhelp/index.html?von_bertalanffy.htm (accessed September 18, 2012.)

48. Making Waves

Savitsky, Daniel. "On the Subject of High-Speed Monohulls." http://legacy.sname.org/newsletter/Savitskyreport.pdf (accessed September 19, 2012).

Simon, Donald C. "Wake Patterns." www.steelnavy.com/WavePatterns.htm (accessed September 19, 2012).

49. A Drop in the Bucket

Edgeworth, R., B. J. Dalton, and T. Parnell. "The Pitch Drop Experiment." *European Journal of Physics* (1984): 198–200.

"Fluids: Kinematic Viscosities." www.engineeringtoolbox.com/kinematic-viscosity-d_397.html (accessed September 25, 2012).

50. Fracking Unbelievable

Jaeger, J. C., N. G. W. Cook, and R. W. Zimmerman. *Fundamentals of Rock Mechanics*. 4th edition. Pp. 412–13. Blackwell Publishing, 1997.

"Oilfield Services: The Unsung Masters of the Oil Industry." *Economist*, July 21, 2012. www.economist.com/node/21559358.

51. Take Two Aspirins and Call Me in the Morning

Craen, Anton J. M., et al. "Placebos and Placebo Effects in Medicine: Historical Overview." *Journal of the Royal Society of Medicine* 92 (1999): 511–15.

Jacobs, B. "Biblical Origins of Placebo." *Journal of the Royal Society of Medicine* 93, no. 4 (April 2000): 213–14.

Thomas, K. B. "General Practice Consultations: Is There Any Point Being Positive?" *British Medical Journal* 294 (1987): 1200–1202.

52. The World's Most Famous Equation

Bodanis, David. *$E = mc^2$: A Biography of the World's Most Famous Equation*. Walker, 2000.

Index

A Brief History of Time (Hawking), ix
AC (alternating current), 14–17
Adler, Mortimer J., x
aging, 136–38
Allen, Woody, 87
Amontons, Guillame, 100–102
Analytical Theory of Heat (Fourier), 105
Annals of Physics, 171
Annie Hall (film), 87
Arago, François, 2, 94
Archimedes, 41–42, 67–69
Aristotle, 83
Arrhenius, Svante, 106
arsenic, 133–35
Asian carp, 153–55

Bangladesh, 37, 133–35
Barry, Dave, 42
Baumgartner, Felix, 130–32
Beck, Glenn, 38
Becquerel, Henri, 118
bell curve. *See* standard normal distribution
bends, the, 50–53
Bernoulli, Daniel, 11–13
Bernoulli, Jacob, 72
Bertalanffy, Ludwig van, 153–55
binary arithmetic, 25–27
Binet, Alfred, 113–15
BMI (body mass index), 22–24
BMR (basal metabolic rate), 141–42
Boole, George, 25–27
Brigham, Carl, 115
Brooklyn Bridge, 50–53
buckling, 90–92
buoyancy principle, 41, 67–69
Buys-Ballot, C.H.D., 19

caisson disease. *See* bends, the
The Canterbury Tales (Chaucer), 167
Celsius, Anders, 56
Challenger (space shuttle), 96–99
Chernobyl, 28
Clark, David, 132
colony collapse disorder, 83–86
conservation of energy, law of, 61–63
Coulomb, Charles Augustin de, 101
Curie, Marie, 118
Curie, Pierre, 118

Darwin, Charles, 116
da Vinci, Leonardo, 9–10, 34, 36, 101
DC (direct current), 14–17
De Architectura (Vitruvius), 68
Deepwater Horizon, 149–52
directional drilling, 164–65
Doppler, Christian Johann, 18–21
Doppler effect, 18–21
Drake, Frank, 127–29

e (mathematical constant), 58, 70–72
$E = mc^2$, ix, 40, 112, 171–73
Edison, Thomas, 14–17, 47
Einstein, Albert: celebrity of, 69, 145; $E = mc^2$, ix, 40, 112, 171–73; proof of Pythagorean theorem, 74; relativity, 3, 63
elastic modulus, 36–37
electromagnetic spectrum, 87–89, 143–45
Electronics Magazine, 32–33
Euler, Leonhard, 58–60, 72, 90
exoskeleton, 90–92
exponential growth, 31–33

Fahrenheit, Daniel Gabriel, 55–56
Faraday, Michael, 94–95
Fechner, Gustav, 9
Fermat, Pierre de, 77–79
Fermi paradox, 128

Feynman, Richard, 99
First law of thermodynamics, 61–63
fluid mechanics, 11–13
Foucault, Jean, 111–12
Fourier, Jean-Baptiste Joseph, 103–6
Fukushima, 28, 123–26

Galileo, 55–56, 68, 110, 112
Galton, Francis, 7
Gardner, Howard, 25
Garfield, James, 74
Gauss, Karl Friedrich, 7, 75
giraffe, 11–13
glass transition temperature, 97–99
golden ratio, 8–10
gravity, 1–3
Great Pyramid, 9
greenhouse effect, 105–6

Hagen-Poiseuille equation, 160–63
Hawking, Stephen, ix
Haygarth, John, 169
Henry, William, 52
Henry's law, 50–53
Herschel, William, 2
Hiero II (king), 67
honeybee, 83–86
Hooke, Robert, 3, 34–36
Hooke's law, 34–36
How to Read a Book (Adler and Van Doren), x
hull speed, 156–59
Humphrey, Nicholas, 80
hydraulic fracturing, 164–66

infrared radiation, 88, 106, 139–42, 143–45
Intel, 32–33
IQ (intelligence quotient), 23, 113–15

Jefferson, Thomas, 168
Jordan, Michael, 23

Keillor, Garrison, 6
Kittinger, Joseph, 132
kudzu, 154–55

Lake Wobegon, 6
LaPlace, Pierre Simon, 7
Lenz, Heinrich, 94
Le Verrier, Urbain, 2–3
Lord Kelvin (William Thomson), x, 26, 101, 105, 116–18

Madoff, Bernie, 107–9
magnetic induction, 93–95
Mainstone, John, 162–63
Mars Climate Orbiter, 65–66
Maxwell, James Clerk, 7
Mead, Carver, 32
Medawar, Peter, 136–37
Mercury, 3
Michelson, Albert, 112
Mitchell, Joni, 39
Moivre, Abraham de, 7
Mona Lisa, 8–9, 58
Moore, Gordon, 32–33
Moore's Law, 32–33
MP3, 104–5
Mythbusters, 69

Neptune, 2–3
Newton, Isaac, 1–3, 36, 43
Newton's law of cooling, 43–45
Newton's laws of motion, 1
nuclear energy, 123–26, 173

Obama, Michelle, 24
Ohm, Georg, 146–48
On Burning Mirrors and Lenses (Ibn Sahl), 81
On Floating Bodies (Archimedes), 68

Parnell, Thomas, 160–62
Parthenon, 9
Perkins tractors, 169
perpetual motion, 61–63
pi (π), 40, 59
placebo effect, 167–70
Planck, Max, 143–45
Ponzi, Charles, 107–9
Principia Mathematica (Newton), 2
psychophysics, 9
Pythagorus, 73–76, 105

Quetelet, Adolphe, 22–23

Red Bull Stratos Project, 130–32
relativity, theories of, 3, 112, 171–73
reverberation time, 120–22
Roebling, Washington, 53
Roentgen, Wilhelm, 118
Rogers Commission, 98–99
Rømer, Ole, 56, 111
Ruth, Babe, 172
Rutherford, Ernest, 118

Sabine, Wallace, 119–22
Sagan, Carl, 128
Sahl, Ibn, 81
SAT, 115
SETI (Search for Extraterrestrial Intelligence), 128
Shannon, Claude, 25–27
Snell-Descartes law of refraction, 80–82
speed of light (*c*), 20, 40, 110–12, 129, 143, 171–73
SPF (sun protection factor), 87–89
standard normal distribution, 4–7
Stefan-Boltzman law, 139–42
Stern, Wilhelm, 113–14
Stewart, Potter, 22
Sullivan, Roy, 148
sunlight, 87–89, 144–45
surface tension, 149–52

Tesla, Nikola, 16
Thomas Aquinas, Saint, 71
Thomson, William. *See* Lord Kelvin
Tournament of Roses Parade, 38
transistor, 31–33
tsunami, 28–30
Turing, Alan, 33

U.S. Patent and Trademark Office, 63
Uranus, 2–3
Ussher, James, 116
UV (ultraviolet), 88–89, 137, 144–45

Van Doren, Charles, x
Van Gogh, Vincent, 58
Vidal, Gore, 113
viscoelasticity, 162–63
Vitruvius, 68
Volta, Allessandro, 47

War of Currents, 14–17
Warhol, Andy, 113
Wells, David, 58–59
Westinghouse, George, 14–17
Wiles, Andrew, 77–79
Williams, Serena, 23
The Wizard of Oz (film), 73

zebra mussel, 155